Numerical Calculus

Differentiation and Integration

by D. James Benton

Copyright © 2018 by D. James Benton, all rights reserved.

Foreword

Before the advent of sophisticated programs capable of performing calculus symbolically, numerical differentiation and integration provided a means of solving seemingly intractable equations. Numerical methods can still be an efficient means of solving many such problems, but the real advantage of *Numerical Calculus* will always be in solving those problems that have no closed-form solution—and these are legion. This book is filled with practical examples, code, and spreadsheets. I trust you will find it useful. I assume that you already have a command of analytical calculus and so I will jump right in to the numerical.

All of the examples contained in this book,
(as well as a lot of free programs) are available at...
https://www.dudleybenton.altervista.org/software/index.html

Programming

Most of the examples in this book are implemented in the C programming language. Perhaps the biggest shortfall of BASIC® is that it has no data statements (e.g., x(1)=0.111 is an *executable* statement, not a *data* statement). As we shall see, data statements are at the very heart of numerical integration. I have included several complete programs in the free on-line archive that illustrate the many uses of numerical integration as well as the constants and code to implement every type and order you could possibly want. You aren't likely to find 4096-pt Gauss Quadrature anywhere else on the Web!

$$\iint\left(\frac{\partial f}{\partial x}-\frac{\partial g}{\partial y}\right)dxdy = \oint(fdx+gdy)$$

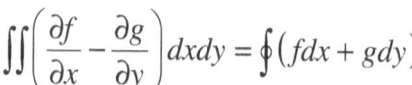

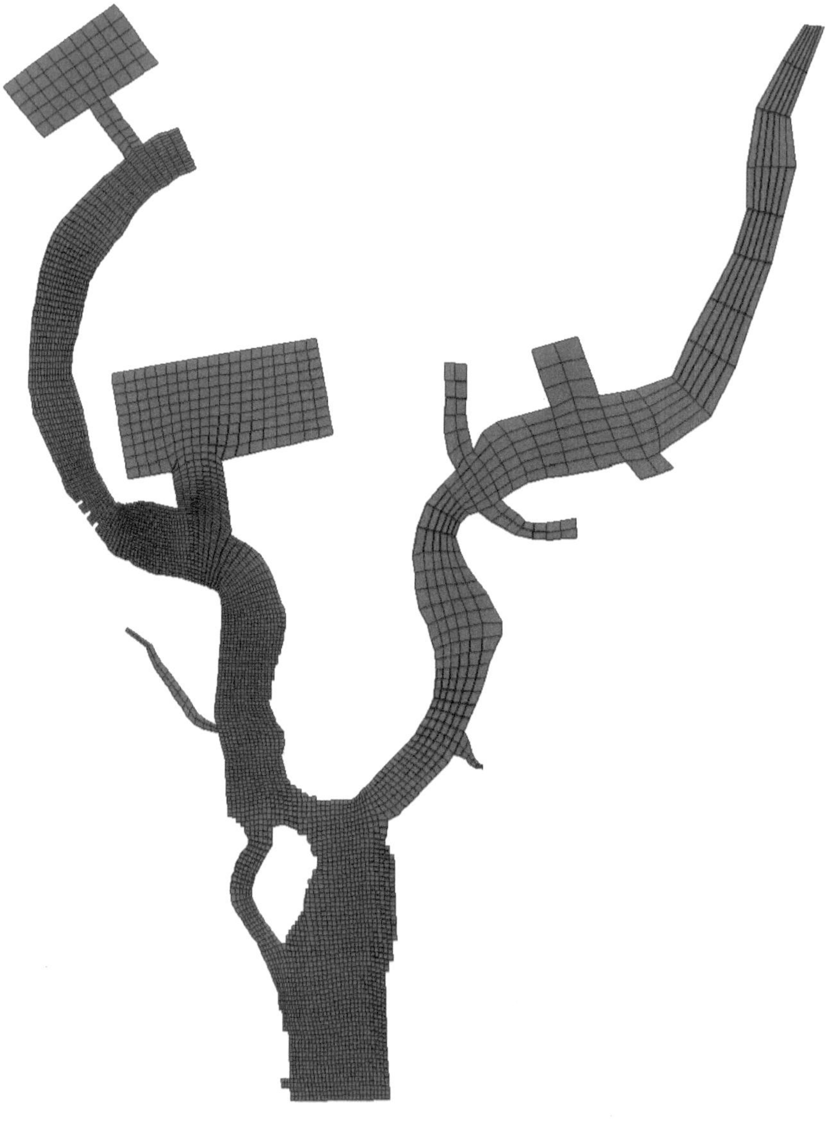

Table of Contents   page

Foreword i
Programming i
Chapter 1. Data Analysis 1
Chapter 2. Newton-Cotes Integration 5
Chapter 3. Gauss Quadrature 11
Chapter 4. Chebyshev & Lobatto Quadrature 17
Chapter 5. Composite Rules 21
Chapter 6. Green's Lemma 25
Chapter 7. 2D and 3D Integrals 29
Chapter 8. Improper Integrals 33
Chapter 9. Applications 39
Appendix A: Data Transformation Program 73
Appendix B. Newton-Cotes Coefficient Program 79
Appendix C. Gauss Quadrature Weights & Abscissas 83
Appendix D. Cooling Tower Demand 99

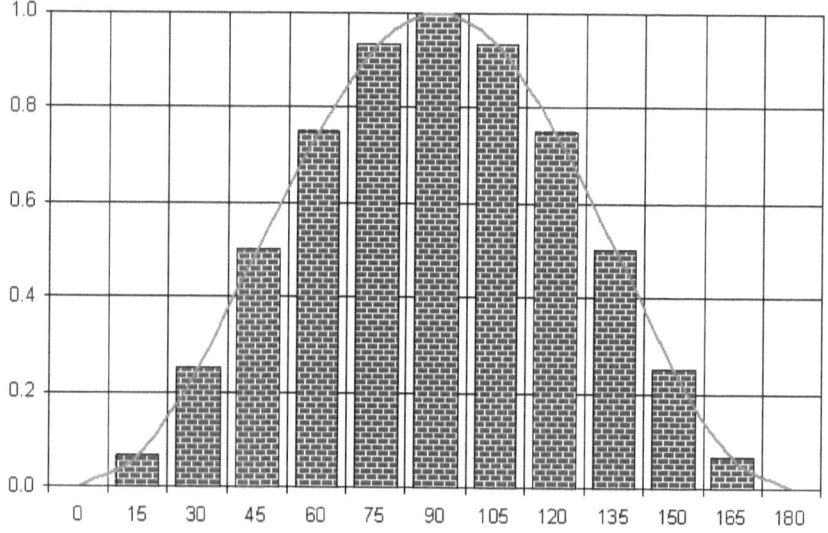
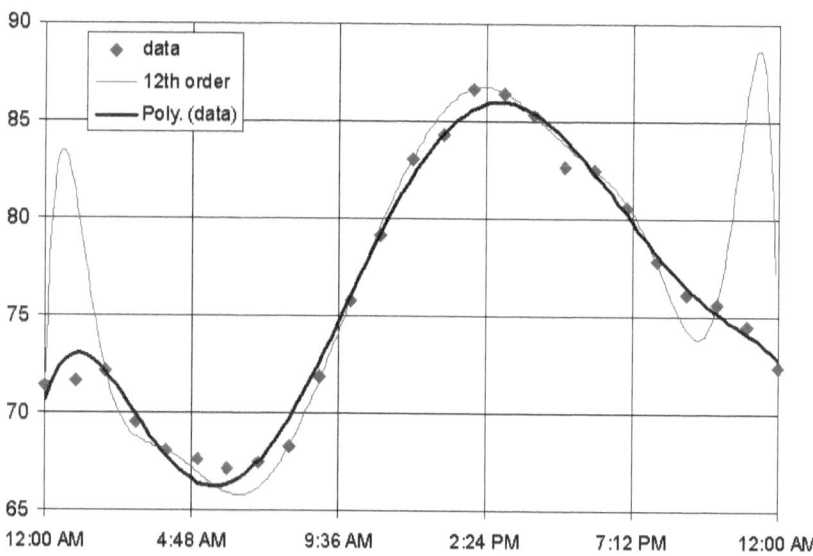

Chapter 1. Data Analysis

The first application of numerical calculus to be considered is data analysis. In this first chapter the focus will be limited to evenly spaced data. The two most common measurements meeting this criterion are time (e.g., 0, 5, 10, 15, 20 seconds) and length (e.g., 0.1, 0.2, 0.3, 0.4 cm). The quantity measured is often temperature (degrees), volts, amps, watts, or grams. The analysis methods discussed in this chapter apply equally well to all of these measurements.

Several things may be extracted from this data that implicitly involve calculus, including: 1) the slope or rate of change, 2) the accumulation or integration, 3) the initial state (backward in time), and 4) the final state (forward in time). An example of the first category would be the rate of temperature change from which might be deduced thermal conductivity or insulating properties. An example of the second category would be the rate of deposition or filling from which we might deduce concentration or flow.

The third and fourth categories are often overlooked but are of significant practical interest. Sometimes it is not practical or even possible to measure the initial state of a system. Starting up an experiment may in itself be a disruption that interferes with initial measurements. It is often not practical to conduct very lengthy experiments. Both of these situations provide motivation for extrapolation, either backward or forward in time—something we can easily do with numerical calculus. This figure illustrates all four categories:

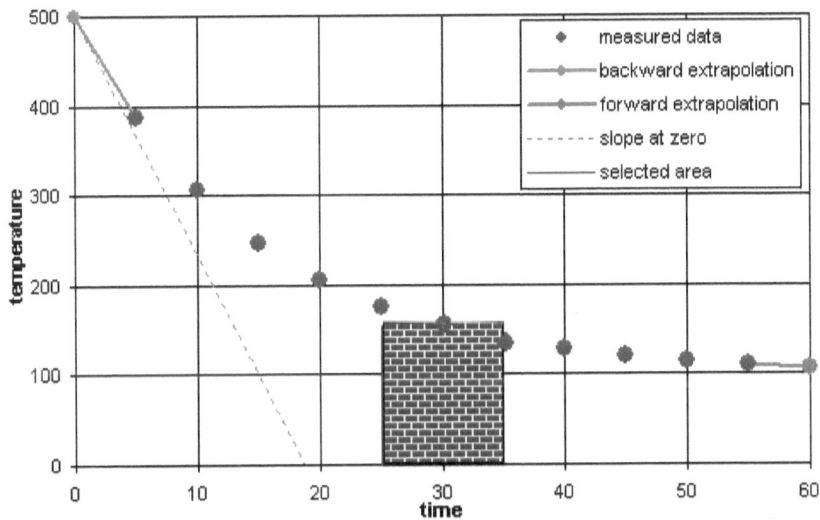

In this case, it's simple to come up with the first order relationships. For the temperature at t=0 sec and t=60 sec:

$$T_0 \approx 2T_5 - T_{10} \tag{1.1}$$

1

$$T_{60} \approx 2T_{55} - T_{50} \qquad (1.2)$$

The slope at t=0 is approximately the same as the slope at t=5, or:

$$\left.\frac{dT}{dt}\right|_{t=0} \approx \frac{T_{10} - T_5}{\Delta t} \qquad (1.3)$$

where $\Delta t = 5$. The selected area is approximately:

$$A_{25-35} \approx T_{30}\, 2\Delta t \qquad (1.4)$$

The second order approximations are a little more complicated:

$$T_0 \approx 3T_5 - 3T_{10} + T_{15} \qquad (1.5)$$

$$T_{60} \approx T_{45} - 3T_{50} + 3T_{55} \qquad (1.6)$$

$$\left.\frac{dT}{dt}\right|_{t=0} \approx \frac{-5T_5 + 8T_{10} - 3T_{15}}{2\Delta t} \qquad (1.7)$$

$$A_{25-35} \approx (T_{25} + T_{35})\Delta t \qquad (1.8)$$

Algebraic expressions for third order approximations are quite tedious to arrive at without a symbolic processor such as Maple®. While third order integration isn't a problem, extrapolation and differentiation are. These can have considerable error and be much less accurate than second and first order. Still, sometimes we would like to know these results. This problem naturally leads to the topic of smoothing.

It is possible to calculate the derivative using 4 points, which would be third order, but also require one degree of smoothing, effectively reducing this estimation to second degree, while achieving greater accuracy than second degree and taking an additional data point into consideration. It is also possible to consider 5 points plus 2 degrees of smoothing and so forth. Extrapolation can also be combined with smoothing and even more points so that many combinations are possible.

While such approximations are often tailored for each specific application, it is possible to implement this process in general. The usefulness of such an approach is readily apparent when applied to repetitive measurements or some sort of production process. Another application for this technique is process control—calculations that are carried out continuously.

The conventional approach for determining these coefficients for combined smoothing, integrating, differentiating, and extrapolating would be to set up a matrix called a Vandermonde and then invert it. The general form of such a matrix is as follows:

$$A = \begin{bmatrix} 1 & X_1 & X_1^2 & X_1^3 \\ 1 & X_2 & X_2^2 & X_2^3 \\ 1 & X_3 & X_3^2 & X_3^3 \\ 1 & X_n & X_n^2 & X_n^3 \end{bmatrix} \qquad (1.9)$$

This is how one typically fits a polygon (i.e., y=a+b*x+c*x²+d*x³) to a series of points and then integrates or differentiates the polygon. This sounds simple enough, but it quickly runs out of steam with large orders (e.g.,x^{50}) due to accumulated round-off errors and limited precision floating-point calculations. Inverting a matrix is always subject to such errors.

Thankfully, there is a completely different approach to solving this problem that doesn't involve inverting a matrix and isn't subject to accumulated round-off error: orthogonal polynomials. Details of this approach are described and the source code listed in Appendix A. That program will be used here to obtain coefficients. For example:

```
transformation coefficients for equally-spaced data
p=points, i=point, d=differentiate (-1 for
   integrate),s=smoothing

p=2,i=0,d=0,s=0
2,-1                       Equation 1.1

p=2,i=0,d=1,s=0
-1,1                       Equation 1.3

p=3,i=0,d=0,s=0
3,-3,1                     Equation 1.5

p=3,i=0,d=1,s=0
-2.5,4,-1.5                Equation 1.7
```

This program accommodates real numbers for parameter i so that you can estimate the value, slope, or integral at locations in between the points, for example midway between points 1 and 2 would be 1.5,as in:

```
transformation coefficients for equally-spaced data
p=points, i=point, d=differentiate (-1 for
   integrate),s=smoothing

p=3,i=1.5,d=0,s=0
0.375,0.75,-0.125          the value at 1.5

p=3,i=1.5,d=1,s=0
-1,1,0                     the derivative at 1.5
```

```
p=3,i=1.5,d=-1,s=0
0.666667,0.416667,-0.0833333
```

This next sequence of coefficients reveals the impact of smoothing along with increasing order in estimating the value at zero (for which we don't have data):

```
transformation coefficients for equally-spaced data
p=points, i=point, d=differentiate (-1 for
    integrate),s=smoothing

p=2,i=0,d=0,s=0
2,-1

p=3,i=0,d=0,s=1
1.33333,0.333333,-0.666667

p=4,i=0,d=0,s=2
1,0.5,0,-0.5

p=5,i=0,d=0,s=3
0.8,0.5,0.2,-0.1,-0.4

p=6,i=0,d=0,s=4
0.666667,0.466667,0.266667,0.0666667,-0.133333,-0.333333

p=7,i=0,d=0,s=5
0.571429,0.428571,0.285714,0.142857,0,-0.142857,-
    0.285714

p=8,i=0,d=0,s=6
0.5,0.392857,0.285714,0.178571,0.0714286,-0.0357143,-
    0.142857,-0.25
```

Notice that each set of coefficients in this set sum to unity. This is always the case with extrapolation, interpolation, and integration. For differentiation the coefficients will always sum to zero.

Further Differentiation

If these coefficients with smoothing aren't adequate for your differentiation needs, I suggest you fit a curve through the data and analytically differentiate it. For more details on curve-fitting, see my book on the subject:

https://www.amazon.com/dp/B01LDUK032

Chapter 2. Newton-Cotes Integration

The earliest method for numerical integration more sophisticated than the Trapezoidal Rule[1] is attributed to Isaac Newton[2] and Roger Coates[3]. This method uses equally-spaced points (i.e., $x_{i+1}-x_i$=constant) and can be represented by the following formula:

$$\int_a^b y\,dx \approx (b-a)\sum_{i=1}^{n} C_i y(x_i)$$
$$x_i = \frac{a(n-i)+b(i-1)}{(n-1)} \tag{2.1}$$

The method works fairly well for a small number of points but falls apart for large values of n. The reason this and some other methods *fall apart* is very important and why this method is presented here. The reasoning behind Newton-Cotes integration can be summarized as follows: If a polygon of order n-1 went through every one of the points (y_i) and that polygon was analytically integrated, what would be the result? The general polygon of order n-1 that goes through each y_i at x_i is the *Lagrange interpolator* and is given by the following formula:

$$p_n(x) = \sum_{i=1}^{n} \frac{y_i \prod_{j=1}^{n}(x-x_j)_{j\neq i}}{\prod_{j=1}^{n}(x_i-x_j)_{j\neq i}} \tag{2.2}$$

This formula can be expanded and integrated analytically in order to obtain the coefficients. The first few sets of coefficients are as follows:

$$n=2: \frac{1}{2}, \frac{1}{2} \tag{2.3}$$

$$n=3: \frac{1}{6}, \frac{2}{3}, \frac{1}{6} \tag{2.4}$$

$$n=4: \frac{1}{8}, \frac{3}{8}, \frac{3}{8}, \frac{1}{8} \tag{2.5}$$

$$n=5: \frac{7}{90}, \frac{32}{90}, \frac{12}{90}, \frac{32}{90}, \frac{7}{90} \tag{2.6}$$

---

[1] The Trapezoidal Rule is so elementary that it won't be covered here.
[2] Sir Isaac Newton 1642-1746: English mathematician, astronomer, theologian, author, and physicist.
[3] Roger Cotes 1682-1716: English mathematician, Fellow of the Royal Society.

Newton-Cotes for n=4 is often called Simpson's Rule[4] and n=5 is often called Boole's Rule[5]. The coefficients can be determined without solving a matrix. A program to do so is listed in Appendix B. The coefficients for orders up to 60 are listed there.

All of the coefficients are positive for orders less than or equal to 8. Beginning at order 9 one or more of the coefficients is negative. As these always sum to unity, this means that one or more of the coefficients must be larger than one for orders above 8. This leads to divergence.

With any such method, we would hope that increasing order would consistently increase accuracy, but this is not the case with this method due to this instability. In order to test the Newton-Cotes method, a simple integral with an analytical solution is chosen:

$$\int_0^\pi \sin(x)^2 = \frac{\pi}{2} \qquad (2.6)$$

The results for orders 3 through 60 are as follows:
```
order  3: 2.094395
order  4: 1.767146
order  5: 1.535890
order  6: 1.551975
order  7: 1.572666
order  8: 1.571918
order  9: 1.570724
order 10: 1.570751
order 11: 1.570798
order 12: 1.570798
order 16: 1.570796
order 20: 1.570796
order 30: 1.570796
order 40: 1.570796
order 50: 1.570797
order 60: 1.565138
exact:    1.570796
```

For this example, we achieve seven digits of accuracy with orders 16 through 40 and then lose accuracy for higher orders. The test program (NCtest.c) may also be found in the examples\ NewtonCotes folder.

Extended Rules

Simpson's Rule (a.k.a. Newton-Cotes for n=4) is often extended in some repeating pattern. We will consider such methods here, as they are of historical interest; however, they are of little value in practice. Far more efficient methods

---

[4] Thomas Simpson (1710-1761): English mathematician.
[5] George Boole (1815-1864): English mathematician, educator, philosopher, and logician.

are available that provide increased accuracy with fewer function evaluations, thus less computing time. We will consider several variants. The first of these might be called the 1-4-2 Rule, as the coefficients are 1 on each end plus alternating 4 and 2 in between. For 5 points this becomes: 1,4,2,4,1 and for 7 points becomes: 1,4,2,4,2,4,1. The second variant might be called the 1-3-2 Rule. For 5 points this is: 1,3,2,3,1 and for 7 points: 1,3,2,3,2,3,1. Divide by the sum in each case to normalize. The code to perform this task is rather simple:

```c
#define _CRT_SECURE_NO_DEPRECATE
#include <stdio.h>
#include <stdlib.h>
#define _USE_MATH_DEFINES
#include <float.h>
#include <math.h>
double f(double x)
   {
   return(pow(sin(x),2));
   }
double Simpson(double f(double x),double a,double b)
   {
   return((b-
   a)*(f(a)+3.*f((a+2.*b)/3.)+3.*f((2.*a+b)/3.)+f(b))/8.
   );
   }
double Simpson132(int n,double f(double x),double
    a,double b)
   {
   int i;
   double r,s;
   if(n<5||n%2==0)
     return(-999.);
   s=f(a)+f(b);
   r=2.;
   for(i=2;i<=n-1;i+=2)
     {
     s+=3.*f((a*(n-i)+b*(i-1))/(n-1))+2.*f((a*(n-i-
     1)+b*i)/(n-1));
     r+=3.+2.;
     }
   return((b-a)*s/r);
   }
double Simpson142(int n,double f(double x),double
    a,double b)
   {
   int i;
   double r,s;
   if(n<5||n%2==0)
     return(-999.);
   s=f(a)+f(b);
```

```
      r=2.;
      for(i=2;i<=n-1;i+=2)
        {
         s+=4.*f((a*(n-i)+b*(i-1))/(n-1))+2.*f((a*(n-i-
         1)+b*i)/(n-1));
         r+=4.+2.;
        }
      return((b-a)*s/r);
      }
    int main(int argc,char**argv,char**envp)
      {
      int n;
      printf("Testing Simpson's Rules\n");
      printf("4 pt: %lf\n",Simpson(f,0.,M_PI));
      for(n=5;n<=61;n+=n>21?4:2)
        printf("%i pt: %lf or %lf\n",n,
           Simpson132(n,f,0.,M_PI),
           Simpson142(n,f,0.,M_PI));
      printf("exact: %lf\n",M_PI/2.);
      return(0);
      }
```

The results for the same integral $(\sin(x)^2)$ are listed below:

```
Testing Simpson's Rules
4 pt: 1.767146
5 pt: 1.308997 or 1.346397
7 pt: 1.385997 or 1.413717
9 pt: 1.427997 or 1.449966
11 pt: 1.454441 or 1.472622
13 pt: 1.472622 or 1.488123
15 pt: 1.485888 or 1.499396
17 pt: 1.495997 or 1.507964
19 pt: 1.503954 or 1.514696
21 pt: 1.510381 or 1.520125
23 pt: 1.515681 or 1.524596
27 pt: 1.523907 or 1.531526
31 pt: 1.529996 or 1.536649
35 pt: 1.534686 or 1.540589
39 pt: 1.538409 or 1.543714
43 pt: 1.541436 or 1.546253
47 pt: 1.543945 or 1.548356
51 pt: 1.546059 or 1.550128
55 pt: 1.547865 or 1.551640
59 pt: 1.549425 or 1.552946
exact: 1.570796
```

The performance is quite poor, which is why nobody uses either one of these methods. There are some diehards that won't let go of Simpson's method and so they came up with yet another variant using subdivision. This algorithm is even easier to implement with a reentrant call: split the interval in half and apply the 1,3,3,1 Rule to each half, making 8 points. You can split each of those intervals in half, making it 16 points, and so forth. Here's the code:

```c
#define _CRT_SECURE_NO_DEPRECATE
#include <stdio.h>
#include <stdlib.h>
#define _USE_MATH_DEFINES
#include <float.h>
#include <math.h>
double f(double x)
   {
   return(pow(sin(x),2));
   }
double SplitSimpson(int n,double f(double x),double
     a,double b)
   {
   if(n)
      return(SplitSimpson(n-
   1,f,a,(a+b)/2.)+SplitSimpson(n-1,f,(a+b)/2.,b));
   return((b-
     a)*(f(a)+3.*f((a+2.*b)/3.)+3.*f((2.*a+b)/3.)+f(b))/8.
     );
   }
int main(int argc,char**argv,char**envp)
   {
   int n;
   printf("Testing Split Simpson's Rule\n");
   for(n=0;n<=5;n++)
      printf("%i pt:
   %lf\n",4<<n,SplitSimpson(n,f,0.,M_PI));
   printf("exact: %lf\n",M_PI/2.);
   return(0);
   }
```

This works much better than the extended Simpson Rules, as you can see:

```
Testing Split Simpson's Rule
4 pt: 1.767146
8 pt: 1.570796
16 pt: 1.570796
32 pt: 1.570796
64 pt: 1.570796
128 pt: 1.570796
exact: 1.570796
```

Chapter 3. Gauss Quadrature

Gauss[6] Quadrature[7] is the premier method for numerical integration. Unlike many other methods (e.g., Newton-Cotes), it just keeps getting better as the number of points increases. Not only does the convergence increase with more points, the rate of convergence also increases—in practice, if not in theory. Gauss Quadrature (GQ) of order n will precisely integrate any polynomial up to order n+1.

The biggest difference between GQ and other methods (with the exception of Chebyshev and Lobatto Quadrature, which we will discuss in the next chapter) is the x-values (i.e., the location of the points) are also selected so as to improve the result. In the context of numerical integration or quadrature, the coefficients are often called *weights* and the points are often called *abscissas*. The Newton-Cotes abscissas were evenly-spaced and the weights were determined by requiring the exact integral of any polynomial up to some order. The same is true for GQ, but the addition of optimal point selection add twice as many conditions per order. This addition eliminates the problem of negative weights and divergence.

Orthogonal Polynomials

The details of orthogonal polynomials are beyond the scope of this work, although these were introduced in Chapter 1. Those polynomials were orthogonal when summed up at evenly-spaced points. The polynomials we're interested in here are orthogonal when integrated over the interval from -1 to +1, or:

$$\int_{-1}^{+1} P_i(x) P_j(x) = \frac{2}{2n+1}, i = j \qquad (3.1)$$
$$\int_{-1}^{+1} P_i(x) P_j(x) = 0, i \neq j$$

The polynomials that satisfy this condition are called Legendre[8]. These can be generated through several relationships, including the following:

$$P_n(x) = \frac{1}{2^n} \sum_{k=0}^{n} B(n,k) B\left(\frac{n+k-1}{n}, n\right) x^k \qquad (3.2)$$

where *B(i,j)* is the binomial coefficient. More often these are generated by the recursion relationship:

---

[6] Carl Friedrich Gauss (1777-1855): German mathematician.
[7] The term *quadrature* comes from making squares as in drawing boxes to implement the Trapezoidal Rule for integration.
[8] Adrien-Marie Legendre (1752-1833): French mathematician.

$$(n+1)P_{n+1}(x) = (2n+1)xP_n(x) - nP_{n-1}(x) \tag{3.3}$$

where $P_0(x)=1$ and $P_1(x)=x$. The first 6 are shown in the following figure:

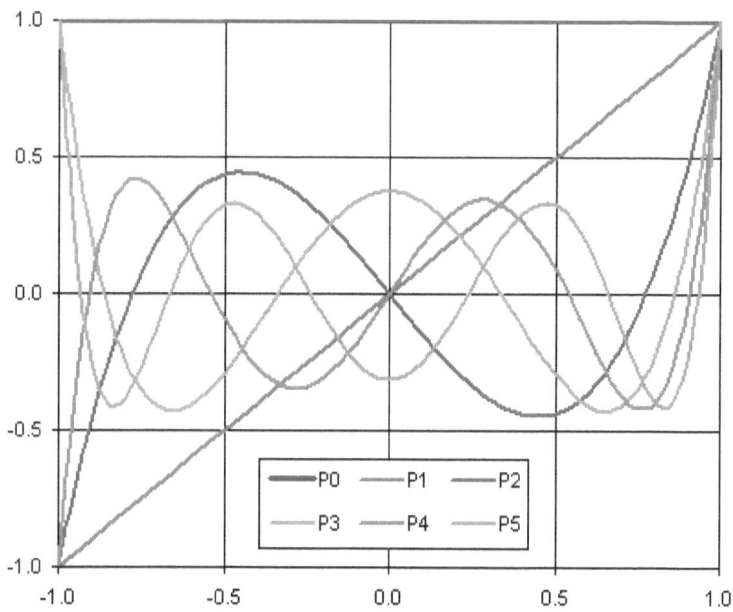

Weights & Abscissas

We could go through the lengthy process of setting up the simultaneous equations to find the n weights and n abscissas such that any polynomial of order up to order 2n would be precisely integrated over the interval from -1 to +1. These simultaneous equations would be nonlinear due to the powers of x_i. Solution of such equations would be tedious and have limited accuracy. There is no point in doing this because the solution may be found from the properties orthogonality of the Legendre polynomials.

There are n roots (locations where the polynomial passes through y=0) for the n+1 order Legendre polynomial. These are the abscissas. If you plug these values into the n order Legendre polynomial, you get the weights. The proof is quite involved and not the primary concern here. The GQ abscissas and weights for orders 3 through 6 are listed here and for orders 7 through 12 are listed in Appendix C. Abscissas and weights for orders up to 4096 may be found in the examples\GaussQuadrature folder in the on-line archive.

```
double A3[]={0.000000000000000,0.774596669241483};
double W3[]={0.888888888888888,0.555555555555556};
double A4[]={0.339981043585,0.861136311594};
double W4[]={0.652145154863,0.347854845137};
```

```
double A5[]={0.000000000000,0.538469310106,
   0.906179845939};
double W5[]={0.568888888889,0.478628670499,
   0.236926885056};
double A6[]={0.238619186083,0.661209386466,
   0.932469514203};
double W6[]={0.467913934573,0.360761573048,
   0.171324492379};
```

The code to implement GQ is very simple. There's a slight difference for odd orders, as the central abscissa is zero and the weight is only used once. The rest are trivial to implement. Here's the function:

```
double GQuad(double*A,double*W,int n,double f(double
   x),double a,double b)
{
int i;
double ax,dx,q;
ax=(a+b)/2.;
dx=(b-a)/2.;
if(n%2)
  {
  q=W[0]*f(ax);
  for(i=1;i<(n+1)/2;i++)
     q+=W[i]*(f(ax-dx*A[i])+f(ax+dx*A[i]));
  }
else
  {
  for(q=i=0;i<n/2;i++)
     q+=W[i]*(f(ax-dx*A[i])+f(ax+dx*A[i]));
  }
return(q*dx);
}
```

You will find a program (GQtest.c) in the examples\GaussQuadrature folder, most of which is data statements. It calculates this same integral $(\sin(x)^2)$ for orders up to 4096. Here are the results:

```
Testing Gauss Quadrature
order 2: 1.19283364802896540
order 3: 1.60606730241801920
order 4: 1.56911897502940610
order 5: 1.57084471252998000
order 6: 1.57079538896380870
order 7: 1.57079633985864310
order 8: 1.57079632665773450
order 9: 1.57079632679478530
order 10: 1.57079632679580520
order 12: 1.57079632679293630
order 16: 1.57079632679489660
order 20: 1.57079632679605700
```

```
order 32:   1.57079632679489660
order 40:   1.57079632679387940
order 64:   1.57079632679489630
order 128:  1.57079632679489630
order 256:  1.57079632679489700
order 512:  1.57079632679489610
order 1024: 1.57079632679489610
order 2048: 1.57079632679489570
order 4096: 1.57079632679489660
exact: 1.57079632679489660
```

Order 16 achieves the exact solution to 18 significant figures! This is why I say GQ is the *premier* method. In this same folder you will find a little program to test GQ (tank.c) by calculating the volume of an ellipsoidal tank and hemispherical end caps. The implementation is simple. Here are the functions to be integrated as before:

```
double HorizontalCylinder(double h)
  {
  double w,x;
  if(h<0.)
    return(0.);
  x=2.*r-h;
  if(x<=0.)
    return(0.);
  w=2.*sqrt(h*x);
  return(w*l);
  }
double HorizontalElipsoidalCap(double h)
  {
  double a,w,x;
  if(h<0.)
    return(0.);
  x=2.*r-h;
  if(x<=0.)
    return(0.);
  w=2.*sqrt(h*x);
  a=b*w/r/2.;
  return(M_PI*a*w/4.);
  }
double GaussQuadrature(double F(double),double a,double
    b)
  {
  int i;
  double cx,dx,q;
  cx=(a+b)/2;
  dx=(b-a)/2;
  for(q=i=0;i<sizeof(A)/sizeof(A[0]);i++)
    q+=W[i]*(F(cx-dx*A[i])+F(cx+dx*A[i]));
  return(q*dx);
```

}

The function calls to perform the integrals also simple:
```
GaussQuadrature(HorizontalCylinder,0.,h)
GaussQuadrature(HorizontalElipsoidalCap,0.,h)
```

20-point GQ is adequate, as illustrated by the following results:

```
horizontal cylinder volume
r l  h   analytic numeric
1 2 0.5    1.228    1.228
1 2 1.3    4.323    4.323
1 4 0.5    2.457    2.457
1 4 1.3    8.647    8.647
1 6 0.5    3.685    3.685
1 6 1.3   12.970   12.970
2 2 0.5    1.813    1.813
2 2 2.1   13.366   13.366
2 2 3.7   24.276   24.277
2 4 0.5    3.626    3.627
2 4 2.1   26.732   26.733
2 4 3.7   48.553   48.554
2 6 0.5    5.440    5.440
2 6 2.1   40.098   40.099
2 6 3.7   72.829   72.831
3 2 0.5    2.251    2.251
3 2 2.9   27.075   27.075
3 2 5.3   52.860   52.861
3 4 0.5    4.502    4.502
3 4 2.9   54.149   54.150
3 4 5.3  105.720  105.723
3 6 0.5    6.752    6.752
3 6 2.9   81.224   81.225
3 6 5.3  158.580  158.584
horizontal elipsoidal cap
r  b   h   analytic numeric
1 0.5 0.5    0.164    0.164
1 0.5 1.3    0.752    0.752
2 0.5 0.5    0.180    0.180
2 0.5 2.1    2.251    2.251
2 0.5 3.7    4.122    4.122
2 1.0 0.5    0.360    0.360
2 1.0 2.1    4.503    4.503
2 1.0 3.7    8.243    8.243
2 1.5 0.5    0.540    0.540
2 1.5 2.1    6.754    6.754
2 1.5 3.7   12.365   12.365
3 0.5 0.5    0.185    0.185
3 0.5 2.9    4.477    4.477
3 0.5 5.3    9.070    9.070
3 1.0 0.5    0.371    0.371
```

```
3 1.0 2.9    8.954    8.954
3 1.0 5.3   18.140   18.140
3 1.5 0.5    0.556    0.556
3 1.5 2.9   13.431   13.431
3 1.5 5.3   27.210   27.210
3 2.0 0.5    0.742    0.742
3 2.0 2.9   17.907   17.907
3 2.0 5.3   36.279   36.279
3 2.5 0.5    0.927    0.927
3 2.5 2.9   22.384   22.384
3 2.5 5.3   45.349   45.349
```

Open vs. Closed Interval

Notice that -1 and +1 are not included among the abscissas of any order GQ, although the end points do get closer as the order increases. Thus GQ is called an *open* interval method. The end points are included in *closed* methods. The Newton-Cotes (NC) Rules presented in Chapter 2 were all closed interval. There are open interval NC Rules, but these weren't covered because they have so little practical value. There are advantages to having open interval methods, as we will see in Chapter 6. Always use GQ when an open interval method is required.

Chapter 4. Chebyshev & Lobatto Quadrature

These two GQ-related methods are worth mentioning. Chebyshev Quadrature (CQ) is similar to GQ except all the weights are the same. Lobatto Quadrature (LQ) is the same as GQ except that -1 and +1 are the endpoints. CQ is an open method and LQ is closed. When an open method is needed, use GQ. When a closed method is needed and accuracy is desired, use LQ. I can't think of any reason to use CQ, although it has been used.[9]

Weights and abscissas for both Chebyshev and Lobatto Quadrature may be found in the *Handbook of Mathematical Functions* by Abramowitz and Stegun.[10] The abscissas and procedure for implementing Chebyshev Quadrature is quite simple:

```
/* Chebyshev abscissas */
double C2[]={0.5773502692};
double C3[]={0.7071067812};
double C4[]={0.1875924741,0.7946544723};
double C5[]={0.3745414096,0.8324974870};
double C6[]={0.2666354015,0.4225186538,0.8662468181};
double C7[]={0.3239118105,0.5296518105,0.8838617008};
double C9[]={0.1679061842,0.5287617831,0.6010186554,
    0.9115893077};
double Chebyshev(double*C,int n,double f(double
    x),double a,double b)
{int i;
double ax,dx,q;
ax=(a+b)/2.;
dx=(b-a)/2.;
if(n%2)
    {
    q=f(ax);
    for(i=0;i<n/2;i++)
       q+=f(ax-dx*C[i])+f(ax+dx*C[i]);
    }
else
    {
    for(q=i=0;i<n/2;i++)
       q+=f(ax-dx*C[i])+f(ax+dx*C[i]);
    }
```

---

[9] Merkel's method for integrating cooling tower demand uses 4-point Chebyshev Quadrature, perhaps for ease of graphical representation or ease of hand calculations in the days before computers. Merkel, F. Verdunstungskulung, V.D.I. Forschungsarbeiteh (Society of German Engineers Technical Journal), No. 275, Berlin, 1925.

[10] Abramowitz, M. and I. A. Stegun, *Handbook of Mathematical Functions* first published by the National Bureau of Standards as Technical Monograph No. 55. This very useful reference may be obtained free on-line as a PDF from several different web sites.

```
    return((b-a)*q/n);
    }
```

The weights, abscissas, and procedure for implementing Lobatto Quadrature is also quite simple:

```
/* Lobatto weights and abscissas */
double A3[]={0.5,1.};
double W3[]={4./3.,1./3.};
double A4[]={0.44721360,1.00000000};
double W4[]={0.83333333,0.16666667};
double A5[]={0.00000000,0.65465367,1.00000000};
double W5[]={0.71111111,0.54444444,0.10000000};
double A6[]={0.28523152,0.76505532,1.00000000};
double W6[]={0.55485838,0.37847496,0.06666667};
double A7[]={0.00000000,0.46884879,0.83022390,
   1.00000000};
double W7[]={0.48761904,0.43174538,0.27682604,
   0.04761904};
double A8[]={0.20929922,0.59170018,0.87174015,
   1.00000000};
double W8[]={0.41245880,0.34112270,0.21070422,
   0.03571428};
double A9[]={0.0000000000,0.3631174638,0.6771862795,
   0.8997579954,1.0000000000};
double W9[]={0.3715192744,0.3464285110,0.2745387126,
   0.1654953616,0.0277777778};
double A10[]={0.1652789577,0.4779249498,0.7387738651,
   0.9195339082,1.0000000000};
double W10[]={0.3275397612,0.2920426836,0.2248893420,
   0.1333059908,0.0222222222};
double Lobatto(double*A,double*W,int n,double f(double
   x),double a,double b)
  {int i;
  double ax,dx,q;
  ax=(a+b)/2.;
  dx=(b-a)/2.;
  if(n%2)
    {
    q=W[0]*f(ax);
    for(i=1;i<(n+1)/2;i++)
      q+=W[i]*(f(ax-dx*A[i])+f(ax+dx*A[i]));
    }
  else
    {
    for(q=i=0;i<n/2;i++)
      q+=W[i]*(f(ax-dx*A[i])+f(ax+dx*A[i]));
    }
  return(q*dx);
  }
```

A program to test each of these (CLtest.c) may be found in the examples\Lobatto folder of the on-line archive. Here are the results:

```
Testing Chebyshev Quadrature
order 2: 1.19283364797927430
order 3: 1.46010768479294240
order 4: 1.59614564019955290
order 5: 1.58292576263350940
order 6: 1.56933737248257190
order 7: 1.56999618095089670
order 9: 1.57083236167838860
Testing Lobatto Quadrature
order 3: 2.09439510239319530
order 4: 1.52507871486619420
order 5: 1.57286574390776800
order 6: 1.57073868336479850
order 7: 1.57079740532856140
order 8: 1.57079633358758320
order 9: 1.57079632717886280
order 10: 1.57079632665996140
exact: 1.57079632679489660
```

The Chebyshev is more often than not superior to Newton-Cotes with the same number of points and the Lobatto is almost as accurate as Gauss Quadrature.

Closed Interval

There are a few problems when the end points are critical and a closed method is needed. In these cases, Lobatto is preferred over Gauss. One such case is heat exchanger analysis. A temperature *pinch* often occurs at the inlet or exit of a heat exchanger. In such cases, the *tightness* of the *pinch* dominates the performance or dictates the size and expense.

Evaporative cooling towers are an example of this situation. The approach to the wet-bulb temperature (which is a type of pinch) drives the cost more than anything else. A close approach tower cost a lot more than a long approach one. It was previously mentioned that Merkel chose the 4-point Chebyshev method to integrate the governing differential equation and produce cooling tower demand curves. Chebyshev is not a closed method and does not include the end points or the pinch. Lobatto would have been a far better choice for this task, as I have demonstrated.[11]

---

[11] Benton, D. J., "Comparison of Methods for Numerical Integration in Computing Cooling Tower Demand Curves," Cooling Technology Institute, Performance Committee Report, Dana Point, California, 1989.

Chapter 5. Composite Rules

We have already considered some composite rules, including the extended and subdivided Simpson. The same type of subdivision with reentrant calls can be used to easily increase the number of points for most any other method. This will usually improve the accuracy, but doesn't necessarily improve the efficiency of the calculation.

Romberg's Method

At this point we mention Romberg's[12] method. The most significant feature of this method is that it provides an estimate of the error, which can be used to decide how many subdivisions are enough. It has very little practical value, as it takes a lot of points and there are so many more efficient methods available. Here is the code if you want to try it out:

```
double Romberg(double F(double),double X1,double X2,int
   n)
{
int i,i1,j,j1,jj;
double dX,E,H,H2,S,X;
static double R[15][15];
dX=X2-X1;
H=dX;
R[0][0]=(F(X1)+F(X2))*H/2.;
if(n<2)
   return(R[0][0]);
jj=1;
i1=0;
for(i=1;i<n;i++)
   {
   S=0.;
   H2=H/2.;
   X=X1-H2;
   for(j=0;j<jj;j++)
      {
      X+=H;
      S+=F(X);
      }
   R[i][0]=(R[i1][0]+H*S)/2.;
   H=H2;
   E=1.;
   j1=0;
   for(j=1;j<=i;j++)
      {
      E*=4.;
      R[i][j]=(E*R[i][j1]-R[i1][j1])/(E-1.);
```

---

[12] Werner Romberg (1909-2003): German mathematician and physicist.

```
        j1=j;
      }
      jj+=jj;
      i1=i;
    }
    return(R[n-1][n-1]);
}
```

In the examples\composite folder you will find a program (testall.c) that compares 75 methods for speed and accuracy, including: trapezoidal, Simpson, Newton-Cotes, Romberg, Chebyshev, Lobatto, and Gauss. The test function is the integral of 1/x from 1 to 2, which is ln(2). Here are the results:

testing various methods for numerical integration				
method	result	error	seconds	digits/sec
exact=ln(10)	2.30258509299404	N/A	N/A	N/A
3-point trapezoidal	3.29318181818181	0.990596725187773	0.000004	1,012
5-point trapezoidal	2.62922118204376	0.326636089049717	0.000007	72,890
10-point trapezoidal	2.37896825396825	0.076383160974208	0.000015	76,794
20-point trapezoidal	2.32071372799289	0.018128634998853	0.000027	65,311
100-point trapezoidal	2.30326634427572	0.000681251281679	0.000037	84,445
1000-point trapezoidal	2.30259178882424	0.000006695830197	0.000470	11,009
10,000-point trapezoidal	2.30258515983237	0.000000066838326	0.005320	1,349
3-point Simpson	2.74090909090909	0.438323997915045	0.000002	228,356
5-point Simpson	2.40790096999774	0.105315877003698	0.000002	391,002
10-point Simpson	2.32057573532623	0.017990642332192	0.000005	372,257
20-point Simpson	2.30399679153549	0.001411698541445	0.000010	285,026
100-point Simpson	2.30258741788446	0.000002324890419	0.000040	140,840
1000-point Simpson	2.30258509321443	0.000000000220390	0.000460	20,993
10,000-point Simpson	2.30258509299405	0.000000000000006	0.005310	2,675
4-point Newton-Cotes	2.56339285714285	0.260807764148811	0.000002	280,896
5-point Newton-Cotes	2.38570042860365	0.083115335609608	0.000003	411,872
6-point Newton-Cotes	2.35981628020375	0.057231187209712	0.000003	396,005
7-point Newton-Cotes	2.32471556804540	0.022130475051354	0.000003	474,436
8-point Newton-Cotes	2.31875667290220	0.016171579908159	0.000004	425,421
9-point Newton-Cotes	2.30946438340678	0.006879290412737	0.000005	459,522
10-point Newton-Cote	2.30777965561224	0.005194562618199	0.000005	442,612
11-point Newton-Cote	2.30492526495466	0.002340171960618	0.000005	491,074
12-point Newton-Cote	2.30438957505264	0.001804482058597	0.000006	445,843
16-point Newton-Cote	2.30283669104050	0.000251598046457	0.000009	404,920
20-point Newton-Cote	2.30262462735199	0.000039534357950	0.000010	440,303
30-point Newton-Cote	2.30258560470063	0.000000511706588	0.000014	461,338
40-point Newton-Cote	2.30258510168987	0.000000008695833	0.000020	403,034
50-point Newton-Cote	2.30258542275379	0.000000329759748	0.000023	283,579
60-point Newton-Cote	2.29959125650675	0.002993836487292	0.000032	78,868

Method	Value	Error	Time	Count
3-point Romberg	2.74090909090909	0.438323997915045	0.000001	243,579
5-point Romberg	2.38570042860365	0.083115335609608	0.000003	398,368
9-point Romberg	2.31362792006895	0.011042827074904	0.000005	391,384
17-point Romberg	2.30341497733484	0.000829884340796	0.000009	346,611
33-point Romberg	2.30261516949073	0.000030076496687	0.000017	271,306
65-point Romberg	2.30258555868970	0.000000465695660	0.000032	197,872
129-point Romberg	2.30258509581234	0.000000002818299	0.000053	160,313
257-point Romberg	2.30258509300029	0.000000000006244	0.000080	140,056
513-point Romberg	2.30258509299405	0.000000000000004	0.000150	96,330
1025-point Romberg	2.30258509299404	0.000000000000002	0.000470	31,384
2049-point Romberg	2.30258509299405	0.000000000000010	0.000940	14,904
4097-point Romberg	2.30258509299404	0.000000000000003	0.002030	7,179
8193-point Romberg	2.30258509299402	0.000000000000021	0.004210	3,247
16,385-point Romberg	2.30258509299402	0.000000000000020	0.008440	1,624
3-point Chebyshev	2.18520609828096	0.117378994713080	0.000002	569,876
4-point Chebyshev	2.25523458707396	0.047350505920079	0.000002	662,338
5-point Chebyshev	2.27069435812671	0.031890734867336	0.000003	570,478
6-point Chebyshev	2.28819920995952	0.014385883034520	0.000003	575,645
7-point Chebyshev	2.29231812702023	0.010266965973816	0.000004	534,425
9-point Chebyshev	2.29895639355763	0.003628699436416	0.000004	553,123
3-point Lobatto	2.74090909090909	0.438323997915045	0.000002	228,356
4-point Lobatto	2.39942749601619	0.096842403022152	0.000002	443,596
5-point Lobatto	2.32615591796764	0.023570824973597	0.000003	600,187
6-point Lobatto	2.30856660691480	0.005981513920759	0.000003	770,705
7-point Lobatto	2.30413500794603	0.001549914951991	0.000004	772,665
8-point Lobatto	2.30299146266670	0.000406369672661	0.000004	826,575
9-point Lobatto	2.30269242163076	0.000107328636723	0.000005	868,281
10-point Lobatto	2.30261357341938	0.000028480425339	0.000005	939,394
2-point Gauss	2.10638297872340	0.196202114270641	0.000001	654,249
3-point Gauss	2.24660974384731	0.055975349146734	0.000002	782,502
4-point Gauss	2.28696952387280	0.015615569121244	0.000002	745,157
5-point Gauss	2.29828311073711	0.004301982256929	0.000002	946,533
6-point Gauss	2.30140808410775	0.001177008886287	0.000003	915,381
7-point Gauss	2.30226434828873	0.000320744705316	0.000004	938,970
8-point Gauss	2.30249790203241	0.000087190961628	0.000004	964,138
9-point Gauss	2.30256142936713	0.000023663626913	0.000004	1,048,542
10-point Gauss	2.30257867788627	0.000006415107776	0.000006	941,194
12-point Gauss	2.30258462257900	0.000000470415039	0.000006	988,675
16-point Gauss	2.30258509048285	0.000000002511189	0.000008	1,021,264
20-point Gauss	2.30258509297903	0.000000000015010	0.000010	1,082,363
40-point Gauss	2.30258509299538	0.000000000001342	0.000020	593,619
96-point Gauss	2.30258509299404	0.000000000000003	0.000040	362,686

2*5-point Gauss	2.30232304578739	0.000262047206652	0.000005	671,554
10*5-point Gauss	2.30258508304717	0.000000009946871	0.000025	320,093
20*5-point Gauss	2.30258509296621	0.000000000027832	0.000053	197,915
100*5-point Gauss	2.30258509299289	0.000000000001151	0.000160	74,619

As mentioned previously, Gauss Quadrature is by far the most efficient with Lobatto and Chebyshev following it. The subdivided methods (e.g., 2*5, 10*5, 25*5, and 100*5 point Gauss) offer no real improvement in performance, even for this simple example. No doubt, other problems would lead to somewhat different results, but I've investigated this and the same trends and generalizations are persistent.

Chapter 6. Green's Lemma

Green's Lemma[13] is usually covered in advanced calculus. This useful theorem transforms an area (2D) integral into a boundary integral (1D). It is often used in developing finite element solutions, which is where I have employed it with good success. The Lemma can be expressed by the following integral:

$$\iint \left(\frac{\partial f}{\partial x} - \frac{\partial g}{\partial y} \right) dxdy = \oint (fdx + gdy) \qquad (6.1)$$

While it can quite complicated to integrate a function over an irregular area (e.g., finite elements), it may not be nearly as complicated to integrate around a boundary. The spanning or basis functions used in finite elements are often simple and can be analytically integrated over one or the other dimension. This step provides the transformation from $\partial f/\partial x \rightarrow f(x,y)$ and $\partial g/\partial y = g(x,y)$.

Integrating around a boundary (most often a polygon) is quite simple. Furthermore, this often doesn't require a high degree of quadrature. In the case of finite element basis functions, 4-point Gauss Quadrature is adequate. The code to implement this is simple:

```
double GreensLemma(double*Xp,double*Yp,int np,double
  U,double V)
{
static double
  A[]={0.339981043584856,0.861136311594053};
static double
  W[]={0.652145154862546,0.347854845137454};
double dX,dY,S,Xa,Xb,X1,X2,Ya,Yb,Y1,Y2;
int i,j;
X2=Xp[np-1];
Y2=Yp[np-1];
for(S=i=0;i<np;i++)
  {
  X1=X2;
  Y1=Y2;
  X2=Xp[i];
  Y2=Yp[i];
  dX=(X2-X1)/2.;
  dY=(Y2-Y1)/2.;
  Xb=(X2+X1)/2.;
  Yb=(Y2+Y1)/2.;
  for(j=0;j<2;j++)
    {
    Xa=pow(Xb-A[j]*dX,U);
```

---

[13] George Green (1729-1841): British mathematical physicist best known for work with electric fields and magnetism.

```
      Ya=pow(Yb-A[j]*dY,V+1.);
      S+=W[j]*dX*Xa*Ya;
      Xa=pow(Xb+A[j]*dX,U);
      Ya=pow(Yb+A[j]*dY,V+1.);
      S+=W[j]*dX*Xa*Ya;
      }
   }
   return(S/(V+1.));
   }
```

In this case we are integrating the basis function X^U*Y^V. Below is a typical finite element grid that we will use as an example:

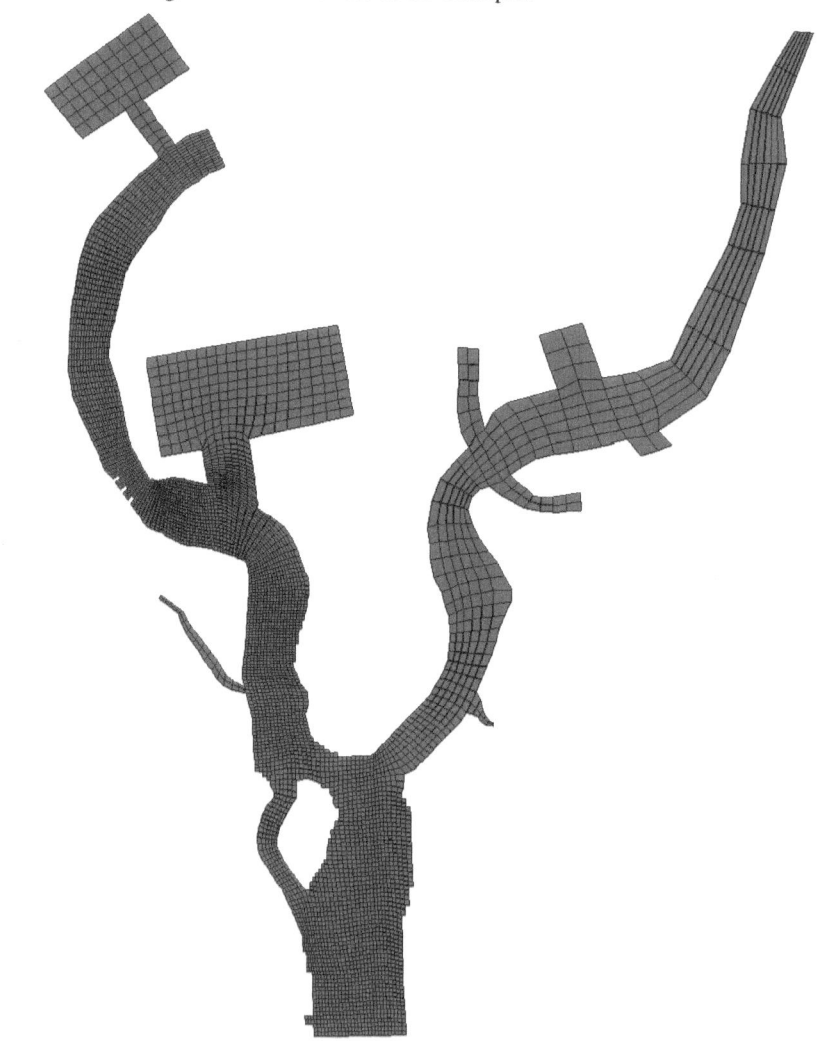

This FEM grid contains 5852 nodes and 5383 elements and may be found in the examples\GreensLemma folder in file grid.2dv. The program to perform the integrations is also in this same folder. Selected results from the calculations are listed below:

```
reading grid: grid.2dv
  5852 nodes
  5383 elements
  U    V    elem    integral
  0.5  0.5    41   -7.07E+013
  0.5  0.5  2318   -8.30E+013
  0.5  0.5   951    2.36E+012
  0.5  0.5  4968   -1.19E+014
  0.5  0.5  3020   -8.20E+013
  0.5  1.0  4958   -1.81E+017
  0.5  1.0   712   -1.08E+017
  0.5  1.0  2443   -1.30E+017
  0.5  1.0    47   -1.19E+017
  0.5  1.0  2932   -1.41E+017
  0.5  1.5   322   -2.01E+020
  0.5  1.5  1230   -2.15E+020
  0.5  1.5  1749   -1.76E+020
  0.5  1.5   678   -1.07E+020
  0.5  1.5  4578   -2.50E+018
  0.5  2.0   491   -3.69E+023
  0.5  2.0  2995   -3.92E+023
  0.5  2.0  1176   -4.08E+023
  0.5  2.0  4827   -3.73E+023
  0.5  2.0    53   -4.91E+023
  1.0  0.5    93   -4.48E+016
  1.0  0.5  3838    2.55E+017
  1.0  0.5  3902    7.09E+016
  1.0  0.5   153   -4.76E+016
  1.0  0.5   292   -4.96E+016
```

The simplest application of Green's Lemma is computing the area of a polygon by integrating around the perimeter. As this is a first order process, the Trapezoidal Rule is adequate and produces exact results. The code is very simple:

```
double polygon_area(double*Xp,double*Yp,int np)
   {
   int i,j;
   double a;
   for(a=j=0,i=np-1;j<np;i=j++)
     a+=(Xp[j]-Xp[i])*(Yp[i]+Yp[j])/2.;
   return(fabs(a));
   }
```

Chapter 7. 2D and 3D Integrals

Sometimes it's necessary to integrate over a 2D or 3D domain. This is easily accomplished with most any of the quadrature formulas. I always use Gauss Quadrature because it's fast and accurate. General 2D integration can be expressed by the following integral:

$$\int_a^b \int_{g(x)}^{h(x)} f(x,y) dx\, dy \tag{7.1}$$

The lower and upper limits of the inner integral are often functions of x and of the outer integral are often constants, but the same logic can be extended to handle functional outer limits of integration. The code to accomplish a 2D integral is quite simple:

```
double GQ2D(double f(double,double),double
    g(double),double h(double),double a,double b)
  {
  int i,j,k;
  double cx,cy,dx,dy,q,w,x,y1,y2;
  cx=(a+b)/2;
  dx=(b-a)/2;
  q=0;
  for(i=0;i<sizeof(A)/sizeof(A[0]);i++)
    {
    for(k=-1;k<=1;k+=2)
      {
      x=cx+k*dx*A[i];
      y1=g(x);
      y2=h(x);
      cy=(y1+y2)/2;
      dy=(y2-y1)/2;
      w=dy*W[i];
      for(j=0;j<sizeof(A)/sizeof(A[0]);j++)
        q+=w*W[j]*(f(x,cy-dy*A[j])+f(x,cy+dy*A[j]));
      }
    }
  return(q*dx);
  }
int n,m;
double f(double x,double y){return(pow(x,m)*pow(y,n));}
double g(double x){return(0.);}
double h(double x){return(M_PI);}
```

As a test case we will consider the following double integral:

$$\int_0^\pi \int_0^\pi x^n y^m dx\, dy = \frac{\pi^{2+n+m}}{(n+1)(m+1)} \tag{7.2}$$

The code (test2D.c) to implement and test this integration, comparing it to the analytical solution, may be found in the examples\2Dplus folder of the on-line archive. The results are as follows:

```
testing 2D integration
n m analytic   numerical
0 0    9.869604    9.869604
0 1   15.503138   15.503138
0 2   32.469697   32.469697
0 3   76.504921   76.504921
0 4  192.277839  192.277839
1 0   15.503138   15.503138
1 1   24.352273   24.352273
1 2   51.003281   51.003281
1 3  120.173649  120.173649
1 4  302.029323  302.029323
2 0   32.469697   32.469697
2 1   51.003281   51.003281
2 2  106.821022  106.821022
2 3  251.691102  251.691102
2 4  632.568734  632.568734
3 0   76.504921   76.504921
3 1  120.173649  120.173649
3 2  251.691102  251.691102
3 3  593.033189  593.033189
3 4 1490.454967 1490.454967
4 0  192.277839  192.277839
4 1  302.029323  302.029323
4 2  632.568734  632.568734
4 3 1490.454967 1490.454967
4 4 3745.921899 3745.921899
```

Recall that n-point Gauss Quadrature will precisely integrate any polynomial up to order n+1. As the highest order in this test is 9 (4+4+1) and 8-point GQ is used, all of the numerical results match the analytical to machine precision. In general, 3D integration can be represented by the following formula:

$$\int_a^b \int_{h1(x,y)}^{h2(x,y)} \int_{g1(x)}^{g2(x)} f(x,y,z) dx dy dz \tag{7.3}$$

The corresponding 3D quadrature is as follows:

```
double GQ3D(double f(double,double,double),double
    g1(double),double g2(double),double
    h1(double,double),double h2(double,double),double
    a,double b)
    {
    int i,j,k,l,m;
    double cx,cy,cz,dx,dy,dz,q,wz,wy,x,y,y1,y2,z1,z2;
    cx=(a+b)/2;
```

```
    dx=(b-a)/2;
    q=0;
    for(i=0;i<sizeof(A)/sizeof(A[0]);i++)
       {
       for(l=-1;l<=1;l+=2)
          {
          x=cx+l*dx*A[i];
          y1=g1(x);
          y2=g2(x);
          cy=(y1+y2)/2;
          dy=(y2-y1)/2;
          wy=dy*W[i];
          for(j=0;j<sizeof(A)/sizeof(A[0]);j++)
             {
             for(m=-1;m<=1;m+=2)
                {
                y=cy+m*dy*A[j];
                z1=h1(x,y);
                z2=h2(x,y);
                cz=(z1+z2)/2;
                dz=(z2-z1)/2;
                wz=dz*W[j];
                for(k=0;k<sizeof(A)/sizeof(A[0]);k++)
                   q+=wy*wz*W[k]*(f(x,y,cz-
    dz*A[k])+f(x,y,cz+dz*A[k]));
                }
             }
          }
       }
    return(q*dx);
    }
int l,m,n;
double f(double x,double y,double
    z){return(pow(x,l)*pow(y,m)*pow(z,l));}
double g1(double x){return(0.);}
double g2(double x){return(M_PI);}
double h1(double x,double y){return(0.);}
double h2(double x,double y){return(M_PI);}
```

The 3D results are similar to the 2D:

```
testing 3D integration
l m n   analytical    numerical
0 0 0     31.00628     31.00628
0 0 1     48.70455     31.00628
0 0 2    102.00656     31.00628
0 0 3    240.34730     31.00628
0 1 0     48.70455     48.70455
0 1 1     76.50492     48.70455
0 1 2    160.23153     48.70455
0 1 3    377.53665     48.70455
```

```
0 2 0    102.00656   102.00656
0 2 1    160.23153   102.00656
0 2 2    335.58814   102.00656
0 2 3    790.71092   102.00656
0 3 0    240.34730   240.34730
0 3 1    377.53665   240.34730
0 3 2    790.71092   240.34730
0 3 3   1863.06871   240.34730
```

Chapter 8. Improper Integrals

Improper integrals have one or more limits that are infinite. There are many such integrals throughout applied mathematics. In general, we will consider the following form:

$$\int_0^\infty f(x)dx \tag{8.1}$$

Other forms (such as from -infinity to +infinity) can easily be recast into this form algebraically, so that there's little point covering those specifically. In Chapter 4 we mentioned the distinction between open and closed quadrature forms. Here is where this important distinction comes into play. We will solve these integrals numerically using *open* quadrature, specifically Gaussian. Consider the following two transformations:

$$\int_0^\infty f(x)dx = \int_0^1 f\left(\frac{1}{y}-1\right)\frac{dy}{y^2} \tag{8.2}$$

$$\int_0^\infty f(x)dx = \int_0^1 f(-\ln(y))\frac{dy}{y} \tag{8.3}$$

We must use open quadrature because we don't want to evaluate the function at either 0 or 1 exactly. With open quadrature, we don't have to because the abscissas never get there. The implementation is simple:

```
double GQ0Ia(double*A,double*W,int n,double f(double))
  {/* using Equation (8.2) */
  int i;
  double q,x;
  for(q=i=0;i<n/2;i++)
    {
    x=(1.-A[i])/2.;
    q+=W[i]*f(1./x-1.)/x/x;
    x=(1.+A[i])/2.;
    q+=W[i]*f(1./x-1.)/x/x;
    }
  return(q/2.);
  }
double GQ0Ib(double*A,double*W,int n,double f(double))
  {/* using Equation (8.3) */
  int i;
  double q,x;
  for(q=i=0;i<n/2;i++)
    {
    x=(1.-A[i])/2.;
    q+=W[i]*f(-log(x))/x;
    x=(1.+A[i])/2.;
    q+=W[i]*f(-log(x))/x;
```

```
        }
    return(q/2.);
    }
double f(double x)
    {
    return(1./(1.+x)/sqrt(x));
    }
```

The code (improper.c) is in the examples\improper folder in the on-line archive. It uses the same weights and abscissas as GQtest.c discussed in Chapter 3. The test problem is:

$$\int_0^\infty \frac{dx}{(1+x)\sqrt{x}} = \pi \qquad (8.4)$$

The results are as follows:

```
Testing Improper Integrals
order      Equation (8.2)       Equation (8.3)
    2  2.44948974278317790  1.79448086891000670
    4  2.75540447558789210  2.06234879066470530
    6  2.87395041793634260  2.17672878406026050
    8  2.93684206475868060  2.24276607150958720
   10  2.97580942795443630  2.28679370330868090
   12  3.00231931310612850  2.31875310599846610
   16  3.03607059068002320  2.36290361690231700
   20  3.05665569109650680  2.39264451109131880
   32  3.08801383542212050  2.44505559952257820
   40  3.09859670208230930  2.46602976782676150
   64  3.11459478032419000  2.50393541705813270
  128  3.12804109039140290  2.54837757177866250
  256  3.13480365100789230  2.58353263249498650
  512  3.13819483897739330  2.61242140200313600
 1024  3.13989291694179280  2.63686388042320230
 2048  3.14074257780349560  2.65800067436659050
 4096  3.14116756381579340  2.67658002342123250
exact  3.14159265358979310
```

Not surprisingly, it takes a lot of points (a high degree of quadrature) to achieve good agreement with the analytical solution. As high order Newton-Cotes Rules (open or closed) are unstable, this precludes using them. We will, however, consider a subdivided approach. The subdivided Simpson code discussed in Chapter 2 can easily be adapted to accommodate the Gaussian abscissas and weights. The function can readily be transformed to implement Equations 8.2 and 8.3 so that the subdivided method can be evaluated. A snippet of the full code (impsplit.c) is listed below:

```
double SplitGauss(double*A,double*W,int n,int m,double
    f(double x),double a,double b)
    {
    int i;
```

```c
  double ax,dx,q;
  if(m)
    return(SplitGauss(A,W,n,m-
   1,f,a,(a+b)/2.)+SplitGauss(A,W,n,m-1,f,(a+b)/2.,b));
  ax=(a+b)/2.;
  dx=(b-a)/2.;
  for(q=i=0;i<n/2;i++)
    q+=W[i]*(f(ax-dx*A[i])+f(ax+dx*A[i]));
  return(q*dx);
  }
double g(double x)
  {
  return(1./(1.+x)/sqrt(x));
  }
double fa(double y)
  {
  return(g(1./y-1.)/y/y);
  }
double fb(double y)
  {
  return(g(-log(y))/y);
  }
int main(int argc,char**argv,char**envp)
  {
  int m;
  printf("Split 8-pt Gauss Quadrature\n");
  printf("point   Eqn(8.2)    Eqn(8.3)\n");
  for(m=0;m<=10;m++)
    printf("%5i %9.7lf
    %9.7lf\n",8<<m,SplitGauss(A8,W8,8,m,fa,0.,1.),
    SplitGauss(A8,W8,8,m,fb,0.,1.));
  printf("Split 16-pt Gauss Quadrature\n");
  printf("point   Eqn(8.2)    Eqn(8.3)\n");
  for(m=0;m<=10;m++)
    printf("%5i %9.7lf
    %9.7lf\n",16<<m,SplitGauss(A16,W16,16,m,fa,0.,1.),
    SplitGauss(A16,W16,16,m,fb,0.,1.));
  printf("Split 32-pt Gauss Quadrature\n");
  printf("point   Eqn(8.2)    Eqn(8.3)\n");
  for(m=0;m<=10;m++)
    printf("%5i %9.7lf
    %9.7lf\n",32<<m,SplitGauss(A32,W32,32,m,fa,0.,1.),
    SplitGauss(A32,W32,32,m,fb,0.,1.));
  printf("exact %9.7lf\n",M_PI);
  return(0);
  }
```

The results are as follows:

```
Split  8-pt Gauss Quadrature
point   Eqn(8.2)   Eqn(8.3)
    8  2.9368421  2.2427661
   16  2.9967531  2.3119032
   32  3.0391547  2.3672321
   64  3.0691506  2.4122695
  128  3.0903658  2.4495589
  256  3.1053688  2.4809500
  512  3.1159782  2.5077959
 1024  3.1234804  2.5310930
 2048  3.1287853  2.5515812
 4096  3.1325365  2.5698141
 8192  3.1351890  2.5862102
Split 16-pt Gauss Quadrature
point   Eqn(8.2)   Eqn(8.3)
   16  3.0360706  2.3629036
   32  3.0669692  2.4087139
   64  3.0888230  2.4465886
  128  3.1042779  2.4784282
  256  3.1152068  2.5056220
  512  3.1229349  2.5291928
 1024  3.1283996  2.5498992
 2048  3.1322637  2.5683088
 4096  3.1349961  2.5848497
 8192  3.1369282  2.5998490
16384  3.1382944  2.6135586
Split 32-pt Gauss Quadrature
point   Eqn(8.2)   Eqn(8.3)
   32  3.0880138  2.4450556
   64  3.1037056  2.4771285
  128  3.1148021  2.5045031
  256  3.1226488  2.5282160
  512  3.1281972  2.5490356
 1024  3.1321207  2.5675365
 2048  3.1348949  2.5841524
 4096  3.1368566  2.5992140
 8192  3.1382438  2.6129760
16384  3.1392246  2.6256373
32768  3.1399182  2.6373547
exact  3.1415927
```

The same pattern holds in that it takes a lot of subdivisions to achieve the accuracy of a higher order method. The first transformation of variables (Equation 8.2) continues to work considerably better than the second (Equation 8.3). The former trend is persistent, but the latter is not always the case. Consider two more transformations:

$$\int_0^\infty f(x)dx = \int_0^1 f\left(\frac{y}{1+y}\right)\frac{dy}{(1+y)^2} \tag{8.5}$$

$$\int_0^\infty f(x)dx = \int_0^{\frac{\pi}{2}} f(\tan(y))\frac{dy}{\cos^2 y} \tag{8.6}$$

The code to implement these is also simple:

```
double GQ0Ic(double*A,double*W,int n,double f(double))
  {
  int i;
  double q,x,y,z;
  for(q=i=0;i<n/2;i++)
     {
     x=(1.-A[i])/2.;
     y=1.+x;
     z=x/y;
     q+=W[i]*f(z)/y/y;
     x=(1.+A[i])/2.;
     y=1.+x;
     z=x/y;
     q+=W[i]*f(z)/y/y;
     }
  return(q);
  }
double GQ0Id(double*A,double*W,int n,double f(double))
  {
  int i;
  double c,d,q,x;
  d=M_PI/4.;
  for(q=i=0;i<n/2;i++)
     {
     x=(1.-A[i])*d;
     c=cos(x);
     q+=W[i]*f(tan(x))/c/c;
     x=(1.+A[i])*d;
     c=cos(x);
     q+=W[i]*f(tan(x))/c/c;
     }
  return(q*d);
  }
```

The program (improper2.c) is compiled and run as before to yield the following results:

```
Testing Improper Integrals
order        Equation 8.5          Equation 8.6
    2  1.71597303267771030  2.22603042920561030
    4  2.06807309586066480  2.65084467337283500
    6  2.19191241793135210  2.80409336963305570
    8  2.25613427256502770  2.88407785764290740
   10  2.29559248245833110  2.93334314579584450
   12  2.32232529582326920  2.96676191112965390
   16  2.35625829270656870  3.00922043971767030
   20  2.37690984517079330  3.03507760555438870
   32  2.40832198945200520  3.07442596377131050
   40  2.41891357351424170  3.08769716442585110
   64  2.43491865828793900  3.10775384473867430
  128  2.44836698031676820  3.12460803595272200
  256  2.45512979549372410  3.13308386900697580
  512  2.45852101548014530  3.13733412052996390
 1024  2.46021909745901100  3.13946234912195220
 2048  2.46106875882368610  3.14052724152694120
 4096  2.46149374489835670  3.14105988255817480
exact  3.14159265358979310
```

Here we see that the transformation in Equation 8.6 is far superior in this case to the one in Equation 8.5.

Chapter 9. Applications

For the remainder of this book we will consider various applications of the principles already presented. Applied mathematics is integral to the sciences, engineering, and even medicine. I trust this assortment of applications will illustrate the power of numerical calculus.

Debye Function

The Debye[14] function or integral arises in the field of thermodynamics and solid-state physics. Peter Debye first encountered it in 1912 while estimating the phonon contribution to the specific heat capacity of a solid. It is given by:

$$D_n(x) = \frac{n}{x^n} \int_0^x \frac{t^n}{e^t - 1} dt \qquad (9.1)$$

It can also be expressed by the following infinite series:

$$D_n(x) = 1 - \frac{n}{2(n+1)} x + n \sum_{k=1}^{\infty} \frac{B_{2k}}{(2k+n)(2k)!} x^{2k} \qquad (9.2)$$

where B_{2k} are the Bernoulli numbers. While it is possible to generate the Bernoulli numbers and sum up this infinite series until the factorial takes over and renders additional terms unnecessary, it is rather tedious. It is much simpler to solve this integral using quadrature and we already have sufficient orders of Gauss Quadrature to do so with precision. The GQ function below has been modified to pass an additional parameter, n (see Eqn. 9.1):

```
double GQuad(double*A,double*W,int m,double
    f(int,double),int n,double a,double b)
{
int i;
double ax,dx,q;
ax=(a+b)/2.;
dx=(b-a)/2.;
for(q=i=0;i<m/2;i++)
   q+=W[i]*(f(n,ax-dx*A[i])+f(n,ax+dx*A[i]));
return(q*dx);
}
double f(int n,double x)
{
return(pow(x,n)/(exp(x)-1.));
}
double Debye(int n,double x)
{
if(x<=0.)
```

---

[14] Peter Debye (1884-1966): Dutch-American physicist and physical chemist; Nobel laureate in Chemistry.

```
        return(1.);
    return(n*GQuad(A4096,W4096,4096,f,n,0.,x)/pow(x,n));
    }
```

The results are more than adequate. This same table below is listed in Chapter 27 of Abramowitz & Stegun. You will also find a snapshot of that page in the applications\Debye folder of the on-line archive.

```
Debye Function
 x       n=1       n=2       n=3       n=4
0.0  1.000000  1.000000  1.000000  1.000000
0.1  0.975278  0.967083  0.963000  0.960555
0.2  0.951111  0.934999  0.926999  0.922221
0.3  0.927498  0.903746  0.891995  0.884994
0.4  0.904437  0.873322  0.857985  0.848871
0.5  0.881927  0.843721  0.824963  0.813846
0.6  0.859964  0.814940  0.792923  0.779911
0.7  0.838545  0.786973  0.761858  0.747057
0.8  0.817665  0.759813  0.731759  0.715275
0.9  0.797320  0.733451  0.702615  0.684551
1.0  0.777505  0.707878  0.674416  0.654874
1.1  0.758213  0.683086  0.647148  0.626228
1.2  0.739438  0.659064  0.620798  0.598598
1.3  0.721173  0.635800  0.595351  0.571967
1.4  0.703412  0.613281  0.570793  0.546317
1.6  0.669366  0.570431  0.524275  0.497882
1.8  0.637235  0.530404  0.481103  0.453132
2.0  0.606947  0.493083  0.441128  0.411893
2.2  0.578427  0.458343  0.404194  0.373984
2.4  0.551596  0.426057  0.370137  0.339218
2.6  0.526375  0.396095  0.338793  0.307405
2.8  0.502682  0.368324  0.309995  0.278355
3.0  0.480435  0.342614  0.283580  0.251879
3.2  0.459555  0.318834  0.259385  0.227792
3.4  0.439962  0.296859  0.237252  0.205915
3.6  0.421580  0.276565  0.217029  0.186075
3.8  0.404332  0.257835  0.198571  0.168107
4.0  0.388148  0.240554  0.181737  0.151855
4.2  0.372958  0.224615  0.166396  0.137169
4.4  0.358696  0.209916  0.152424  0.123913
4.6  0.345301  0.196361  0.139704  0.111956
4.8  0.332713  0.183860  0.128129  0.101180
5.0  0.320876  0.172329  0.117597  0.091471
5.5  0.294240  0.147243  0.095241  0.071228
6.0  0.271260  0.126669  0.077581  0.055677
6.5  0.251331  0.109727  0.063604  0.043730
7.0  0.233948  0.095707  0.052506  0.034541
7.5  0.218698  0.084039  0.043655  0.027453
8.0  0.205239  0.074269  0.036560  0.021968
8.5  0.193294  0.066036  0.030840  0.017702
```

```
9.0  0.182633  0.059053  0.026200  0.014368
9.5  0.173068  0.053092  0.022411  0.011747
10.0 0.164443  0.047971  0.019296  0.009674
```

Diffusion through a Granulated Media

While there is an analytical solution for diffusion through a sphere of a given radius, there is no closed-form solution for one grain among many, approximating a granulated media. Derivation for the analytical solution listed below for a single grain size is beyond the scope of this book.

$$C(r,t) = \frac{C_s}{2}\left[erfc\left(\frac{r-\frac{Vt}{R}-\frac{L}{2}}{2\sqrt{\frac{Dt}{R}}}\right) - erfc\left(\frac{r-\frac{Vt}{R}+\frac{L}{2}}{2\sqrt{\frac{Dt}{R}}}\right)\right] + \frac{(C_0-C_s)}{2}\left[erfc\left(\frac{r-\frac{Vt}{R}-\frac{L}{2}}{2\sqrt{\frac{Dt}{R}}}\right) - erfc\left(\frac{r-\frac{Vt}{R}+\frac{L}{2}}{2\sqrt{\frac{Dt}{R}}}\right)\right] \quad (9.3)$$

Numerical integration is used to find a solution for a distribution of grain sizes. We use Gauss Quadrature to integrate from $\mu-\pi\sigma$ to $\mu+\pi\sigma$, where μ is the mean and σ is the standard deviation of the grain size. In order to perform these calculations, it's necessary to pass a group of parameters to the quadrature function. This is done by defining a structure and passing a pointer to it, as illustrated below:

```
typedef struct{
  double Gm;
  double sG;
  double Co;
  double D;
  double r;
  double t;
  }DIFU;

double GQuad(double*A,double*W,int n,double
    f(DIFU*,double),DIFU*difu,double a,double b)
{
int i;
double ax,dx,q;
ax=(a+b)/2.;
dx=(b-a)/2.;
for(q=i=0;i<n/2;i++)
  q+=W[i]*(f(difu,ax-dx*A[i])+f(difu,ax+dx*A[i]));
return(q*dx);
```

```
  }

double Concentration(double Co,double D,double R,double
  r,double t)
  {
  double s;
  s=2.*sqrt(D*t)/R;
  t*=(1.131302673065950*s+0.2330227286503510)*s+1.;
  r*=(0.404882980327693*s-0.0539281549540029)*s+1.;
  return(Co*(erfc((r-R)/2./sqrt(D*t))-
    erfc((r+R)/2./sqrt(D*t)))/2.);
  }

double Distribution(double Gm,double sG,double G)
  {
  double g;
  g=(G-Gm)/sG;
  return(exp(-g*g/2.)/sG/sqrt(2.*M_PI));
  }

double Contribution(DIFU*difu,double G)
  {
  return(Distribution(difu->Gm,difu-
    >sG,G)*Concentration(difu->Co,difu-
    >D,pow(10.,G)/20.,difu->r,difu->t));
  }

int main(int argc,char**argv,char**envp)
  {
  double Rm,Rx;
  DIFU difu;
  FILE*fp;
  difu.Co=1.;
  difu.D=1E-5;
  difu.Gm=-0.53;
  difu.sG=0.79;
  Rm=pow(10.,difu.Gm)/20.;
  Rx=Rm*3.;
  fp=fopen("diffusion.csv","wt");
  for(difu.r=0;difu.r<Rx;difu.r+=Rx/100)
    {
    fprintf(fp,"%lG",difu.r);
    if(fabs(difu.r)<=Rm)
      fprintf(fp,",1");
    else
      fprintf(fp,",0");
    for(difu.t=0.001;difu.t<101.;difu.t*=10.)
```

```
        fprintf(fp,",%1G",GQuad(A4096,W4096,4096,Contribution
           ,&difu,difu.Gm-difu.sG*M_PI,difu.Gm+difu.sG*M_PI));
        fprintf(fp,"\n");
     }
     fclose(fp);
     return(0);
}
```

All of the associated files may be found in the applications\diffusion folder. The results are shown in the following graph:

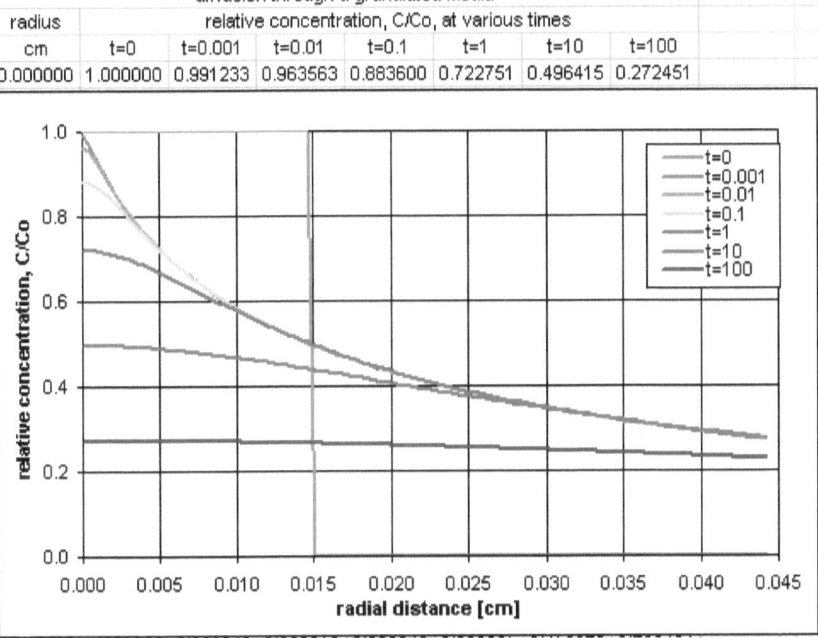

Integral within an Integral

A different solution to this diffusion problem leads to a solution without a closed form, requiring numerical integration. Application of that solution over a range of grain sizes requires another integration, making this an integration within an integration. Again, parameters are passed, but this time instead of creating a type and passing a pointer, the parameters are simply passed through the integrator functions as additional arguments. The code (source.c) and associated files may be found in the applications\diffusion folder.

```
     double Spherical(double Co,double D,double R,double
        r,double t)
     {
```

```c
  double s;
  if(t<FLT_EPSILON)
    {
    if(r<-R)
      return(0.);
    if(r>R)
      return(0.);
    return(Co);
    }
  s=2*sqrt(D*t)/R;
  t*=(1.131302673065950*s+0.2330227286503510)*s+1.;
  r*=(0.404882980327693*s-0.0539281549540029)*s+1.;
  return(Co*(erfc((r-R)/2./sqrt(D*t))-
    erfc((r+R)/2./sqrt(D*t)))/2.);
  }
double OneGrain(double R,double Z,double X,double t)
  {
  return(Spherical(Co,D,R,R,t/Retardation-(Z-L-X)/U)
        -Spherical(Co,D,R,R,t-(Z-L-X)/U*Retardation));
  }
double DistributedGrains(double Z,double X,double t)
  {
  int i;
  double aG,dG,g,G,G1,G2,P,Q,R;
  G1=Gm-sG*M_PI;
  G2=Gm+sG*M_PI;
  dG=(G2-G1)/2.;
  aG=(G1+G2)/2.;
  for(Q=i=0;i<48;i++)
    {
    G=aG-dG*A96[i];
    R=pow(10.,G)/20.;
    g=(G-Gm)/sG;
    P=exp(-g*g/2.)/sG/sqrt(2.*M_PI);
    Q+=P*W96[i]*OneGrain(R,Z,X,t);
    G=aG+dG*A96[i];
    R=pow(10.,G)/20.;
    g=(G-Gm)/sG;
    P=exp(-g*g/2.)/sG/sqrt(2.*M_PI);
    Q+=P*W96[i]*OneGrain(R,Z,X,t);
    }
  return(Q*dG);
  }
double IntegrateOverSourceZone(double Z,double t)
  {
  int i;
  double Q;
  for(Q=i=0;i<48;i++)
    {
```

```
      Q+=W96[i]*DistributedGrains(Z,(1.-A96[i])/2.,t);
      Q+=W96[i]*DistributedGrains(Z,(1.+A96[i])/2.,t);
      }
   return(Q/2);
   }
```

The double integral is utilized as in the following for three different distances:

```
IntegrateOverSourceZone(40.*2.54*12.,t)
IntegrateOverSourceZone(48.*2.54*12.,t)
IntegrateOverSourceZone(50.*2.54*12.,t)
```

The resulting concentration profiles are shown in the figure below:

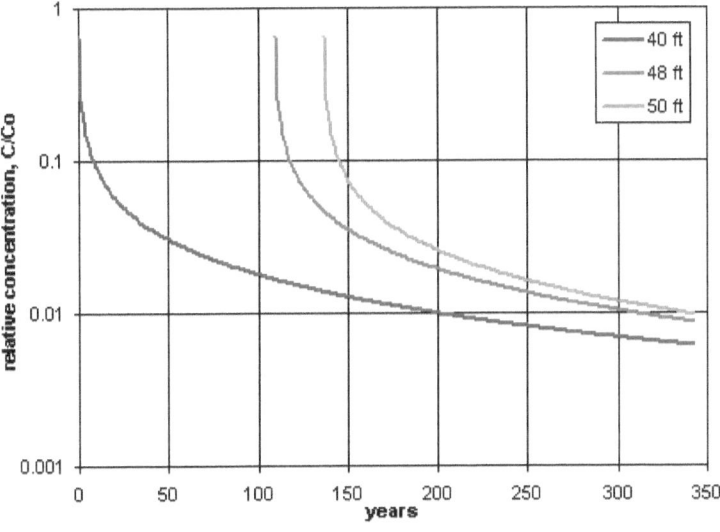

Hybrid Analytical/Numerical Solution

One of the critical outcomes of contaminant transport analyses is breakthrough time, that is, answering the question, "When will the schmutz escape from the containment barrier or when will it arrive at a certain location?" The classical approach (pure diffusive spreading) greatly underestimates this time, as does the zero (i.e., infinitesimal) grain size solution. The finite grain size solution has no closed form and requires integration:

```
double Slab( /* diffusion through a slab */
   double D,  /* diffusion coefficient [ft²/year] */
   double L,  /* thickness of slab [ft] */
   double X,  /* distance from center of slab [ft] */
   double T)  /* time [yr] */
   {
   if(T<FLT_EPSILON)
```

```
    {
    if(X<-L/2.)
      return(0.);
    if(X>L/2.)
      return(0.);
    return(1.);
    }
  return((erfc((X-L/2.)/2./sqrt(D*T))
        -erfc((X+L/2.)/2./sqrt(D*T)))/2.);
  }
double Sphere(/* diffusion through a sphere */
  double D, /* diffusion coefficient [ftý/year] */
  double G, /* radius of grain [ft] */
  double R, /* radial distance from center of grain [ft]
    */
  double T) /* time [yr] */
  {
  double S;
  if(T<FLT_EPSILON)
    {
    if(R<-G)
      return(0.);
    if(R>G)
      return(0.);
    return(1.);
    }
  S=2.*sqrt(D*T)/G;
  T*=(1.131302673065950*S+0.2330227286503510)*S+1.;
  R*=(0.404882980327693*S-0.0539281549540029)*S+1.;
  return((erfc((R-G)/2./sqrt(D*T))
        -erfc((R+G)/2./sqrt(D*T)))/2.);
  }
double Lumped(/* lumped system model */
  double Dh, /* diffusion coefficient [ftý/year] */
  double Ra, /* advective retardation coefficient */
  double Rd, /* diffusive retardation coefficient */
  double Vs, /* seepage velocity [ft/yr] */
  double Lc, /* contaminated length [ft] */
  double X, /* distance from center of contamination
    [ft] */
  double T) /* time [yr] */
  {
  X-=Vs*T/Ra;
  return(Slab(Dh,Lc,X,T/Rd));
  }
double Granular(/* single-grain spherical model */
  double Dh, /* diffusion coefficient [ftý/year] */
  double Ra, /* advective retardation coefficient */
  double Rg, /* radius of grain [ft] */
```

```
  double Vs, /* seepage velocity [ft/yr] */
  double Lc, /* contaminated length [ft] */
  double X, /* distance from center of contamination
    [ft] */
  double T) /* time [yr] */
  {
  double Rm,S;
  X-=Vs*T/Ra;
  Rm=sqrt(Rg*Lc/2.);
  S=2.*M_PI*sqrt(Dh*T/Ra)/Rm;
  return(Slab(Dh,Lc,X,T/Ra)*exp(-S*S));
  }
double Distributed(/* multi-grain spherical model */
  double Dh, /* diffusion coefficient [ftý/year] */
  double Ra, /* advective retardation coefficient */
  double Gm, /* log10(mean radius of grain [ft]) */
  double sG, /* standard deviation of Gm */
  double Vs, /* seepage velocity [ft/yr] */
  double Lc, /* contaminated length [ft] */
  double X, /* distance from center of contamination
    [ft] */
  double T) /* time [yr] */
  {
  int i;
  double aG,dG,g,G,G1,G2,P,Q,Rg;
  G1=Gm-sG*M_PI;
  G2=Gm+sG*M_PI;
  dG=(G2-G1)/2.;
  aG=(G1+G2)/2.;
  for(Q=i=0;i<48;i++)
     {
     G=aG-dG*A96[i];
     Rg=pow(10.,G);
     g=(G-Gm)/sG;
     P=exp(-g*g/2)/sG/sqrt(2.*M_PI);
     Q+=P*W96[i]*Granular(Dh,Ra,Rg,Vs,Lc,X,T);
     G=aG+dG*A96[i];
     Rg=pow(10.,G);
     g=(G-Gm)/sG;
     P=exp(-g*g/2.)/sG/sqrt(2.*M_PI);
     Q+=P*W96[i]*Granular(Dh,Ra,Rg,Vs,Lc,X,T);
     }
   return(Q*dG);
   }
double Classical(double x,double t)
   {
   return(Cp/2*erfc((x-Vs*t/Ra)/2/sqrt(Da*t/Ra)));
   }
double InfiniteGrain(double x,double t)
```

```
    {
    if(Co>Cs)
      return(Cs*Lumped(Da,Ra,1,Vs,Lc,x+Lc/2,t)
        +(Co-Cs)*Lumped(Dd,Rd,1,Vs,Lc,x+Lc/2,t));
    else
      return(Co*Lumped(Da,Ra,1,Vs,Lc,x+Lc/2,t));
    }
  double FiniteGrain(double x,double t)
    {
    if(Co>Cs)
      return(Cs*Distributed(Da,Ra,Gm,sG,Vs,Lc,x+Lc/2,t)
        +(Co-Cs)*Distributed(Dd,Rd,Gm,sG,Vs,Lc,x+Lc/2,t));
    else
      return(Co*Distributed(Da,Ra,Gm,sG,Vs,Lc,x+Lc/2,t));
    }
```

The code (breakthrough.c) and associated files may also be found in the applications\diffusion folder. The breakthrough curves are shown in the following figure:

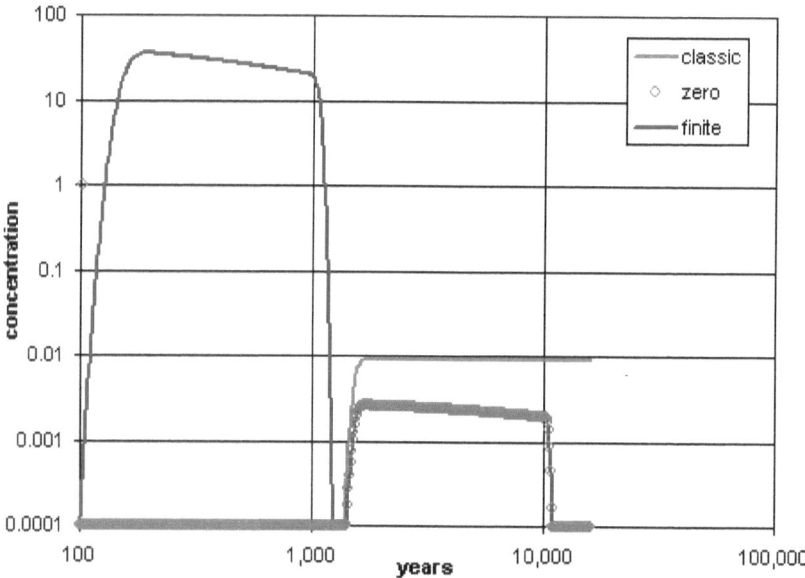

The zero grain size solution (green circles) matches the second part of the finite grain size solution (blue curve), but not the first, which is a lot bigger and a lot sooner! The classic approach (red curve) misses the mark by a mile.

Boundary Element Method

The boundary element method uses Green's Lemma to transform an area integral into a boundary one. Results of numerical boundary integrations go into a matrix solver (simultaneous linear equations) that produces solutions to partial differential equations (PDEs) over an irregular domain. The most common PDE solved in this manner is Laplace's[15], which appears in field theory (gravity, magnetism), invicid fluid flow, and steady-state heat conduction.

There are 3 different integrals: 1) along a boundary adjacent to the current node, 2) along a boundary not directly connected to the current node, and 3) inside the boundary. These 3 functions are all integrated using the same 96-pt Gauss Quadrature:

```
void AdjacentBoundaryIntegral(double X1,double Y1,double
   X2,double Y2,double*G1,double*G2)
   {
   double A,dS;
   dS=hypot(X2-X1,Y2-Y1);
   if(dS<DBL_EPSILON)
     Abort(__LINE__,"adjacent boundary integral
     encountered coincident points");
   A=log(dS);
   *G1=dS*(1.5-A)/2.;
   *G2=dS*(0.5-A)/2.;
   }
void RemoteBoundaryIntegral(double Xp,double Yp,double
   X1,double Y1,double X2,double
   Y2,double*H1,double*H2,double*G1,double*G2)
   {
   int i,n;
   double
    Ai,aX,aY,bX,bY,D,dS,H21,H22,G21,G22,P1,P2,Q,R,T,U,Wi,
    X,Y;
   bX=(X2+X1)/2.;
   bY=(Y2+Y1)/2.;
   aX=(X2-X1)/2.;
   aY=(Y2-Y1)/2.;
   dS=hypot(aX,aY);
   if(fabs(aX)>dS/1000.)
      {
      T=aY/aX;
      D=fabs(T*Xp-Yp+Y1-T*X1)/sqrt(T*T+1.);
      }
   else
      D=fabs(Xp-X1);
   if((X1-Xp)*(Y2-Yp)-(X2-Xp)*(Y1-Yp)<0.)
      D=-D;
   H21=0.;
   H22=0.;
```

---

[15] Pierre-Simon the marquis de Laplace (1749-1847): French mathematician, physicist, and astronomer.

```
      G21=0.;
      G22=0.;
      n=96;
      for(i=0;i<n;i++)
        {
        if(i<n/2)
           {
           Wi= W96[n/2-1-i];
           Ai=-A96[n/2-1-i];
           }
        else
           {
           Wi= W96[i-n/2];
           Ai= A96[i-n/2];
           }
        P1=(1.-Ai)/2.;
        P2=(1.+Ai)/2.;
        X=bX+aX*Ai;
        Y=bY+aY*Ai;
        R=hypot(X-Xp,Y-Yp);
        if(R<DBL_EPSILON)
           Abort(__LINE__,"remote boundary integral
        encountered coincident points");
        U=-log(R);
        Q=-D/(R*R);
        H21+=Wi*P1*Q;
        H22+=Wi*P2*Q;
        G21+=Wi*P1*U;
        G22+=Wi*P2*U;
        }
      *H1=H21*dS;
      *H2=H22*dS;
      *G1=G21*dS;
      *G2=G22*dS;
      }
   int InternalIntegral(double Xp,double Yp,double
      X1,double Y1,double X2,double Y2,

        double*H1o,double*H2o,double*G1o,double*G2o,double*H1
        x,double*H2x,

        double*H1y,double*H2y,double*G1x,double*G2x,double*G1
        y,double*G2y)
      {
      int i,n;
      double
        Ai,aX,aY,bX,bY,D,dS,dX,dY,P1,P2,Q,Qx,Qy,R,Ri,S,T,U,Ux
        ,Uy,Wi,X,Xi,Y,Yi;
      double H21o,H22o;
```

```
double H21x,H22x;
double H21y,H22y;
double G21o,G22o;
double G21x,G22x;
double G21y,G22y;
aX=(X2-X1)/2.;
aY=(Y2-Y1)/2.;
bX=(X2+X1)/2.;
bY=(Y2+Y1)/2.;
dS=hypot(aX,aY);
if(fabs(aX)>dS/1000.)
   {
   T=aY/aX;
   S=sqrt(T*T+1.);
   D=(T*Xp-Yp+Y1-T*X1)/S;
   if(D>=0.)
      {
      dX=T/S;
      dY=-1./S;
      }
   else
      {
      D=-D;
      dX=-T/S;
      dY=1./S;
      }
   }
else
   {
   D=Xp-X1;
   dY=0.;
   if(D>=0.)
      dX=1.;
   else
      {
      D=-D;
      dX--1.;
      }
   }
if((X1-Xp)*(Y2-Yp)<(X2-Xp)*(Y1-Yp))
   {
   D=-D;
   dX=-dX;
   dY=-dY;
   }
H21o=H22o=H21x=H22x=H21y=H22y=0.;
G21o=G22o=G21x=G22x=G21y=G22y=0.;
n=96;
for(i=0;i<n;i++)
```

```
      {
      if(i<n/2)
        {
        Wi= W96[n/2-1-i];
        Ai=-A96[n/2-1-i];
        }
      else
        {
        Wi= W96[i-n/2];
        Ai= A96[i-n/2];
        }
      P1=(1.-Ai)/2.;
      P2=(1.+Ai)/2.;
      X=bX+aX*Ai;
      Y=bY+aY*Ai;
      Xi=X-Xp;
      Yi=Y-Yp;
      Ri=Xi*Xi+Yi*Yi;
      if(Ri<An/1000.)
        return(1);
      R=sqrt(Ri);
      U=-log(R);
      Ux=Xi/Ri;
      Uy=Yi/Ri;
      Q=-D/(R*R);
      Qx=(-2.*D*Xi/Ri-dX)/Ri;
      Qy=(-2.*D*Yi/Ri-dY)/Ri;
      H21o+=Wi*P1*Q;
      H22o+=Wi*P2*Q;
      H21x+=Wi*P1*Qx;
      H22x+=Wi*P2*Qx;
      H21y+=Wi*P1*Qy;
      H22y+=Wi*P2*Qy;
      G21o+=Wi*P1*U;
      G22o+=Wi*P2*U;
      G21x+=Wi*P1*Ux;
      G22x+=Wi*P2*Ux;
      G21y+=Wi*P1*Uy;
      G22y+=Wi*P2*Uy;
      }
  *H1o=H21o*dS;
  *H2o=H22o*dS;
  *H1x=H21x*dS;
  *H2x=H22x*dS;
  *H1y=H21y*dS;
  *H2y=H22y*dS;
  *G1o=G21o*dS;
  *G2o=G22o*dS;
  *G1x=G21x*dS;
```

```
    *G2x=G22x*dS;
    *G1y=G21y*dS;
    *G2y=G22y*dS;
    return(0);
    }
```

The first two integrals are needed to set up the problem and solve for the potential at the boundary points; whereas, the third is needed to evaluate the potential inside the boundary. The results are quite interesting. This first example is for a river bend. Here are the velocity vectors:

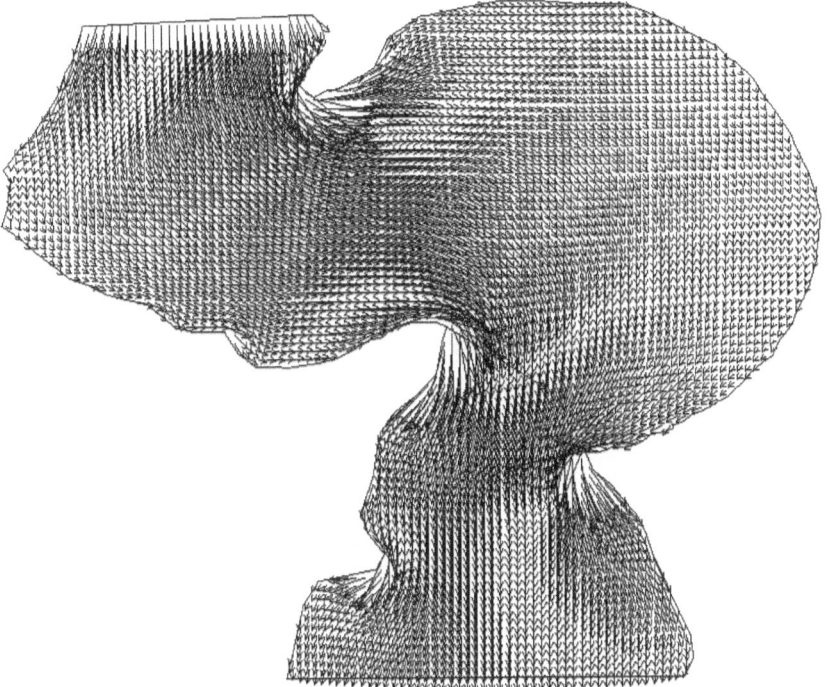

This next figure shows the potential or stream lines:

This next example is for flow in a lake around two islands. First, the velocity vectors:

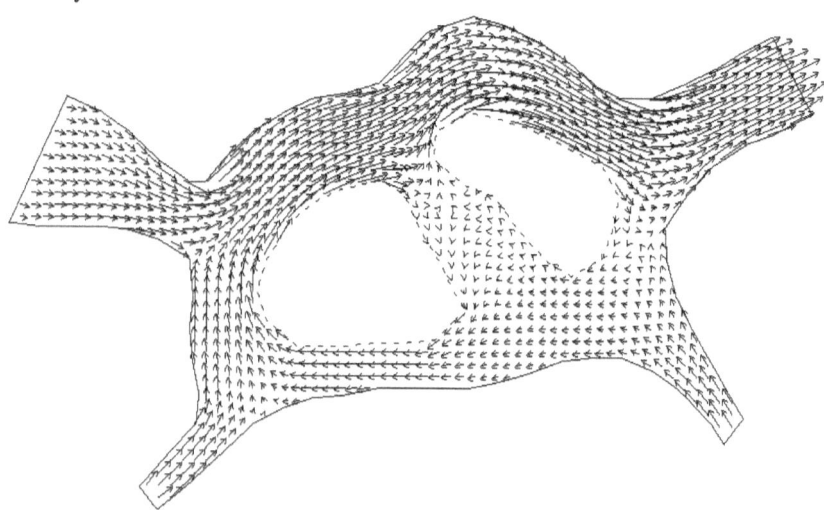

Then the streamlines:

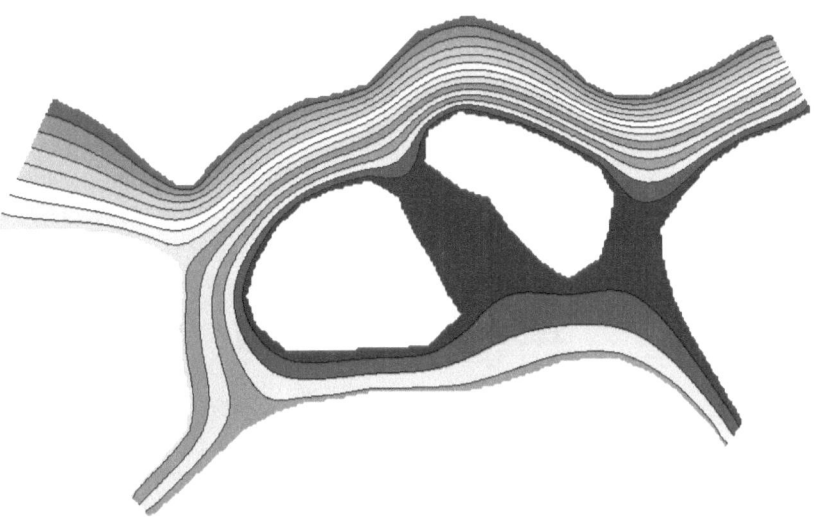

This third example is flow through a natural draft cooling tower:

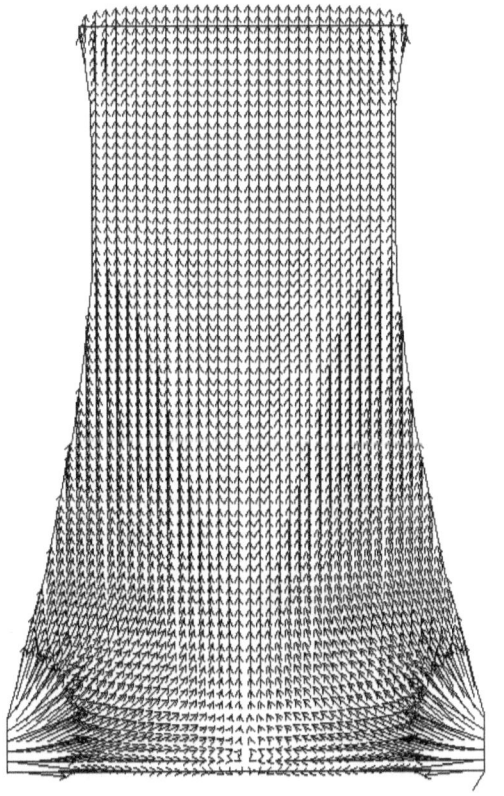

The code (pflow.c), input files, and output are in the applications\pflow folder in the on-line archive. This is a very useful tool and quick to set up. The results are displayed with my general-purpose 2D/3D graphics program, TP2, which is also available free at the same web site.

Heat Exchangers

Heat exchangers are sized according to how demanding the requirements are for the transfer of heat—not just how much heat is transferred, but how close the hot and cold side temperatures approach. A *close approach* heat exchanger is much more expensive than a *long approach* one because it takes that much more material to construct and effort to fabricate. In the case of constant thermal properties, the heat transfer problem has a closed form solution: the log-mean temperature difference. Solutions with variable properties are beyond the scope of this book and readers are directed to my work on that subject:

https://www.amazon.com/dp/B078HM16GV

Within the context of numerical calculus we will only consider constant properties. A spreadsheet (heater.xls) is provided in the applications\heater folder that illustrates the analytical (closed-form) solution along with two numerical ones: 5-pt Lobatto and 4-pt Chebyshev. In this case, the closed-form solution is exact and the numerical ones are approximate. The results are shown in the following figure for typical inputs:

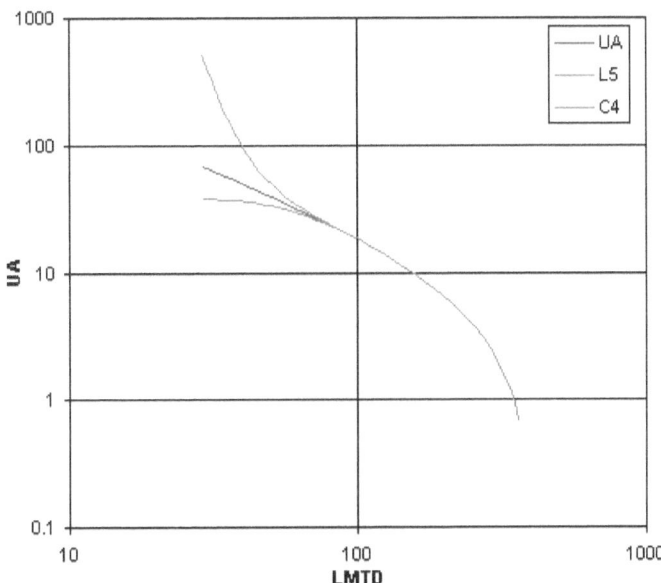

Note that the two numerical solutions (red and green curves) diverge and in opposite directions compared to the analytical solution (blue curve). The 5-pt Lobatto solution (red curve) contains the end points and diverges high (above the analytical) and the 4-pt Chebyshev solution (green curve) does not contain the end points and diverges low (below the analytical). This is important to know when evaluating the integral with variable properties and there is no analytical solution. Some improvement can be gained by averaging the two numerical solutions in the case of variable properties, although using a higher order method yields a superior result in every case that I have ever been able to check.

Cooling Tower Demand Curves

The derivation of cooling tower demand curves is beyond the scope of this book, but is covered thoroughly in Reference 11. Originally, Merkel chose to use the 4-pt Chebyshev method, which is an open method and doesn't include the end points. This works adequately for many problems, but not for *close approach* or high performance designs. An evaporative cooling tower is a type of heat exchanger and the concept of *pinch points* applies here as well. For this type of system, pinches will always occur at the air inlet/water outlet.

The following code calculates the demand using the Merkel (4-pt Chebyshev) and the Exact (10-pt Lobatto) methods:

```
double KMerkel(double Twb,double Ran,double App,double
   LG)
{
int i;
double dHa,Ha,Hai,Hw,KaV,Tco,Tw;
static double a[4]={0.10,0.40,0.60,0.90};
static double w[4]={0.25,0.25,0.25,0.25};
Tco=Twb+App;
Hai=fHtwb(user.baro,Twb);
dHa=Ran*LG;
KaV=0;
for(i=0;i<4;i++)
   {
   Ha=Hai+a[i]*dHa;
   Tw=Tco+a[i]*Ran;
   Hw=fHtwb(user.baro,Tw);
   if(Hw<=Ha)
      return(999.9);
   KaV=KaV+w[i]/(Hw-Ha);
   }
KaV=Ran*KaV;
return(KaV);
}
double KExact(double Twb,double Ran,double App,double
   LG)
{
int i;
double
   Ha,Hai,Hao,KaV,Ta,Tai,Tao,Tao1,Tao2,Tco,Tho,Tw,W,Wai,
   Wao,Ws;
static double a[10]={0.000000,0.040233,0.130613,
   0.261038,0.417361,0.582639,0.738962,0.869387,
   0.959767,1.00000};
static double w[10]={0.011111,0.066653,0.112445,
   0.146021,0.163770,0.163770,0.146021,0.112445,
   0.066653,0.011111};
Tco=Twb+App;
Tho=Tco+Ran;
Tai=Twb;
Hai=fHtwb(user.baro,Tai);
Wai=fWsat(user.baro,Tai);
Tao1=Twb;
Tao2=Tho;
for(i=0;i<32;i++)
   {
   Tao=(Tao1+Tao2)/2.;
   Hao=fHtwb(user.baro,Tao);
```

```
      Wao=fWsat(user.baro,Tao);
      if(Hao<Hai+(LG*Ran+(Wao-Wai)*(Tco-32.018)))
         Tao1=Tao;
      else
         Tao2=Tao;
      }
   KaV=0;
   for(i=0;i<10;i++)
      {
      Ha=Hao+a[i]*(Hai-Hao);
      Ta=fTwbh(user.baro,Ha);
      W=fWsat(user.baro,Ta);
      Tw=((LG*Tho-(Wao-W)*32.018)-Hao+Ha)/(LG-Wao+W);
      Ws=fWsat(user.baro,Tw);
      if(Tw<=Ta)
         return(999.9);
      KaV=KaV+w[i]/(Lewis*0.2406*(Tw-Ta)
      +(1061.39+0.427933*Ta)*(Ws-W)/(1+W));
      }
   KaV=Ran*KaV;
   return(KaV);
   }
```

I have provided the entire code, not only these calculations, but also the Windows® interface as well. You will find all of the files, resources, and compiler commands (_compile.bat) in the applications\KaVL folder. For clarity of comparison, the abscissas for both the Chebyshev and Lobatto integrations have been modified so that they work over the interval 0 to 1 instead of -1 to +1.

The are some additional features in this program you might find useful, including log-linear regression model and associated statistics plus copying data or a graph onto the clipboard. The program will plot performance data (in red) on top of the demand curves (in black). This is how you design a cooling tower: match supply and demand. Supply here is the cooling capacity of the packing (plastic stuff) and demand is the cooling capacity. Where the two meet (i.e., red crosses black) is where the cooling tower will operate. I have included data files for 24 common types of packing (*.?FT). The beginning of the file name is the packing type and the end is the depth in feet. This is the way these devices are typically designed. Ordinarily, you would design a cooling tower to operate in the middle of the middle of the red circles.

The program draws the curves on the unusual green log-log graph paper format that has been used for many decades in the cooling tower industry. Run the program, specify an operating condition and formulation plus optional data file and press the OK button. Select File/Export to copy the graph or numbers to the clipboard. Ctl-V will paste it into any Microsoft® Office® document, as illustrated in this next figure:

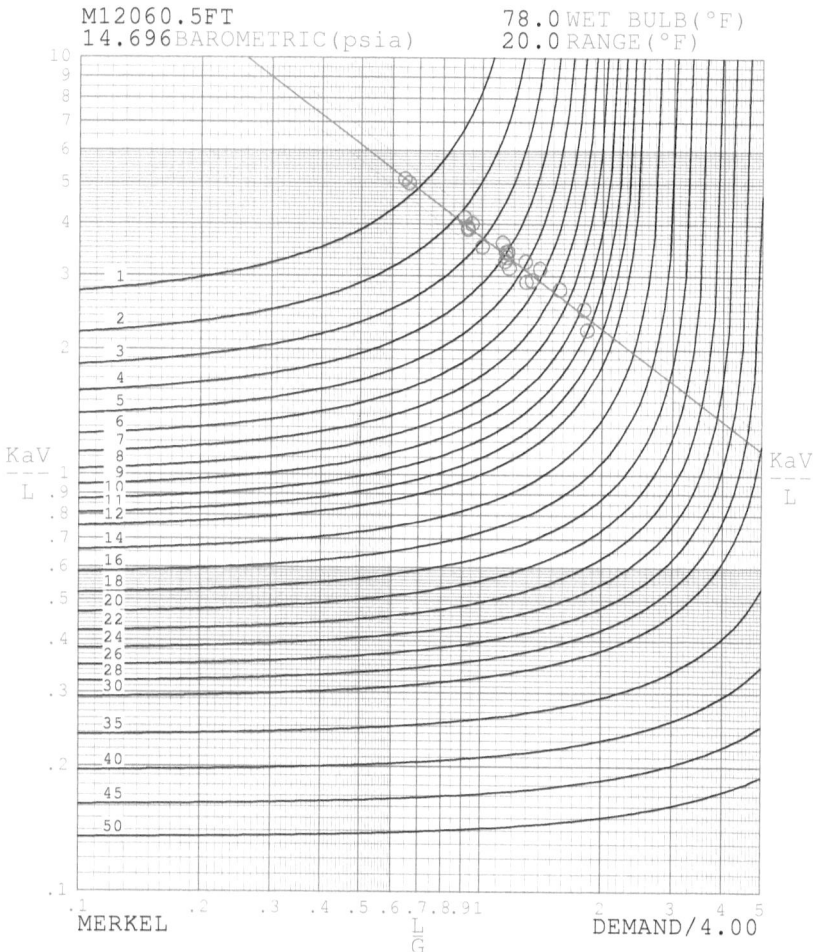

The program works in both English and SI units. The preceding graph is dimensionless but the temperature inputs have dimensions. The results are slightly different for the Exact method using 10-pt Lobatto, as illustrated in this next figure:

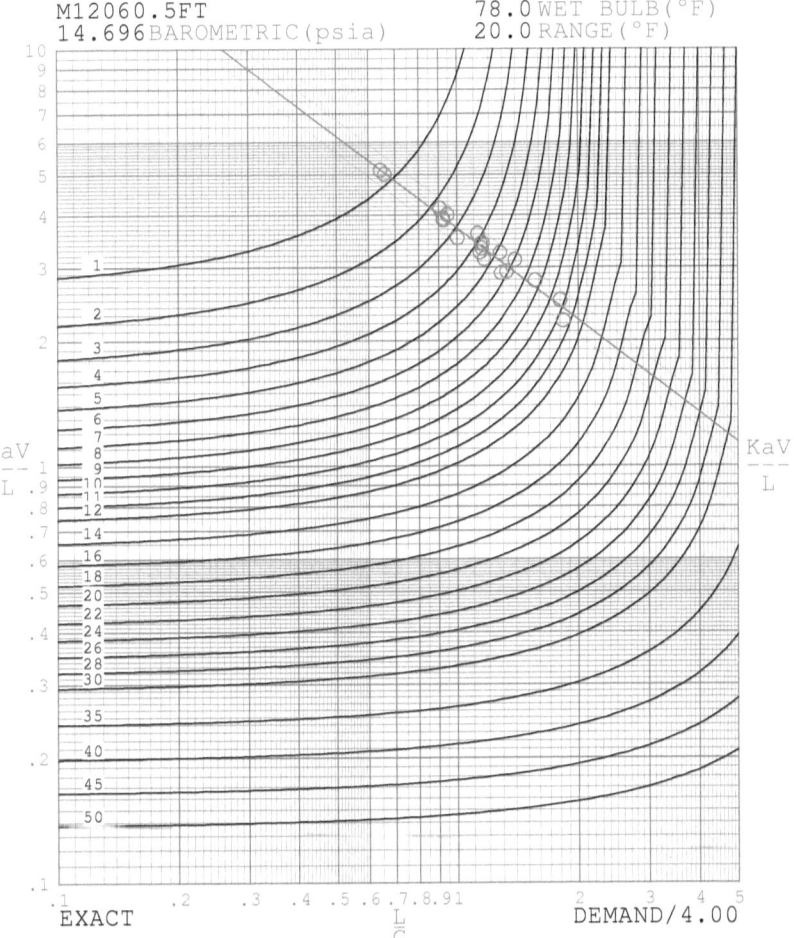

For these operating conditions, the two calculations are not much different. This next figure compares the two integration methods for the most extreme case (1° approach), which would not be a practical design:

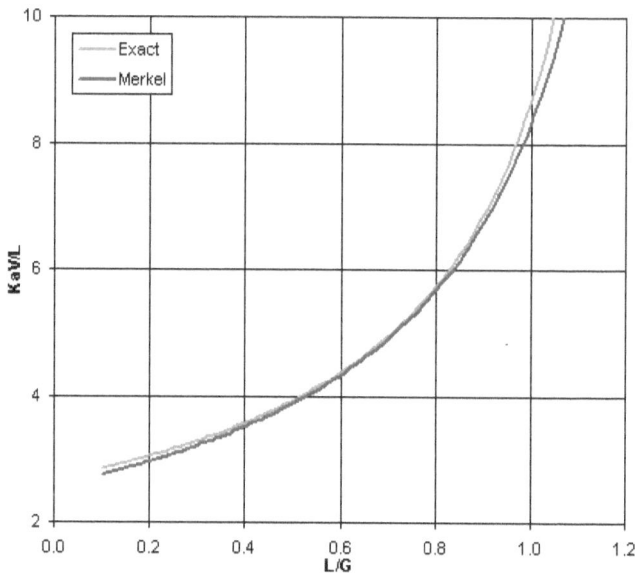

A topic that's not covered in this book—partial differential equations—is included in this example. Merkel's equation only works for counterflow cooling towers. It doesn't work for crossflow. Cooling towers are designed with a crossflow arrangement for a variety of reasons, including the avoidance of fouling or plugging. This program solves the crossflow partial differential equation in 2D using the 4-th order Runge-Kutta method. The program also includes moist air properties, which are essential to the calculations.

Laplace Transforms

The Laplace transform is defined by the following integral:

$$F(s) = \int_0^\infty e^{-st} f(t) dt \qquad (9.4)$$

Tables abound for such transforms having a closed-form solution; however, there are many more functions for which no analytical solution has been found. In such cases, it is possible to evaluate this integral numerically. A simple modification of the first two of four formulas used in Chapter 8 will perform this special improper integral.

```
double GQ0Ia(double*A,double*W,int n,double
   f(int,double),int m,double s)
  {
  int i;
  double q,t,x;
  for(q=i=0;i<n/2;i++)
    {
    x=(1.-A[i])/2.;
    t=1./x-1.;
    q+=W[i]*exp(-s*t)*f(m,t)/x/x;
    x=(1.+A[i])/2.;
    t=1./x-1.;
    q+=W[i]*exp(-s*t)*f(m,t)/x/x;
    }
  return(q/2.);
  }
double GQ0Ib(double*A,double*W,int n,double
   f(int,double),int m,double s)
  {
  int i;
  double q,t,x;
  for(q=i=0;i<n/2;i++)
    {
    x=(1.-A[i])/2.;
    t=-log(x);
    q+=W[i]*exp(-s*t)*f(m,t)/x;
    x=(1.+A[i])/2.;
    t=-log(x);
    q+=W[i]*exp(-s*t)*f(m,t)/x;
    }
  return(q/2.);
  }
```

To check the implementation, we will first evaluate simple integrals for which there is a known solution, specifically:

$$L\{t^m\} = \frac{\Gamma(m+1)}{s^{m+1}} \tag{9.5}$$

Typical program output is listed below:

```
Laplace Transform by Gauss Quadrature
m  s    result1   result2   exact
1 0.5   4.00000   3.99552   4.00000
1 1.0   1.00000   1.00000   1.00000
1 1.5   0.44444   0.44444   0.44444
1 2.0   0.25000   0.25000   0.25000
1 2.5   0.16000   0.16000   0.16000
1 3.0   0.11111   0.11111   0.11111
1 3.5   0.08163   0.08163   0.08163
1 4.0   0.06250   0.06250   0.06250
```

```
1 4.5   0.04938    0.04938    0.04938
1 5.0   0.04000    0.04000    0.04000
2 0.5  16.00000   15.90589   16.00000
2 1.0   2.00000    2.00000    2.00000
2 1.5   0.59259    0.59259    0.59259
2 2.0   0.25000    0.25000    0.25000
2 2.5   0.12800    0.12800    0.12800
2 3.0   0.07407    0.07407    0.07407
2 3.5   0.04665    0.04665    0.04665
2 4.0   0.03125    0.03125    0.03125
2 4.5   0.02195    0.02195    0.02195
2 5.0   0.01600    0.01600    0.01600
3 0.5  96.00000   94.01740   96.00000
3 1.0   6.00000    5.99996    6.00000
3 1.5   1.18519    1.18519    1.18519
3 2.0   0.37500    0.37500    0.37500
3 2.5   0.15360    0.15360    0.15360
3 3.0   0.07407    0.07407    0.07407
3 3.5   0.03998    0.03998    0.03998
3 4.0   0.02344    0.02344    0.02344
3 4.5   0.01463    0.01463    0.01463
3 5.0   0.00960    0.00960    0.00960
4 1.0  24.00000   23.99904   24.00000
4 1.5   3.16049    3.16049    3.16049
4 2.0   0.75000    0.75000    0.75000
4 2.5   0.24576    0.24576    0.24576
4 3.0   0.09877    0.09877    0.09877
4 3.5   0.04570    0.04570    0.04570
4 4.0   0.02344    0.02344    0.02344
4 4.5   0.01301    0.01301    0.01301
4 5.0   0.00768    0.00768    0.00768
```

The numerical results are quite accurate even up to the 4th power. A Laplace Transform that arises in molecular dynamics is the following function:

$$f(t) = t^n \mathrm{erfc}\left(\frac{\sqrt{t}}{2}\right) \qquad (9.6)$$

The error function and its complement can be approximated by:

```
double erf(double x)
  {
  double e,q,t,y;
  if(fabs(x)<FLT_EPSILON)
    return(0.);
  if(x>6.)
    return(1.);
  y=fabs(x);
  t=1./(1.+0.3275911*y);
```

```
  q=((((10.061405429*t-10.453152027)*t+10.421413741)*t-
    0.284496736)*t+0.254829592)*t;
  e=1.-q/exp(x*x);
  if(x<0)
    return(-e);
  return(e);
  }
#define erfc(x) (1.-erf(x))
double f(int m,double t)
  {
  return(pow(t,m)*erfc(sqrt(t)/2.));
  }
```

The results are difficult to verify, but reasonable:

```
Laplace Transform by Gauss Quadrature
m  s    result1   result2
0 0.5   7.80140   7.80140
0 1.0   5.19106   5.19106
0 1.5   3.93109   3.93109
0 2.0   3.17930   3.17930
0 2.5   2.67662   2.67662
0 3.0   2.31544   2.31544
0 3.5   2.04272   2.04272
0 4.0   1.82913   1.82913
0 4.5   1.65711   1.65711
0 5.0   1.51546   1.51546
1 0.5   8.10293   8.10290
1 1.0   3.38576   3.38576
1 1.5   1.88014   1.88014
1 2.0   1.20388   1.20388
1 2.5   0.84020   0.84020
1 3.0   0.62132   0.62132
1 3.5   0.47899   0.47899
1 4.0   0.38106   0.38106
1 4.5   0.31069   0.31069
1 5.0   0.25837   0.25837
2 0.5  18.18250  18.10180
2 1.0   4.69377   4.69377
2 1.5   1.89431   1.89431
2 2.0   0.95471   0.95471
2 2.5   0.55002   0.55002
2 3.0   0.34657   0.34657
2 3.5   0.23287   0.23287
2 4.0   0.16424   0.16424
2 4.5   0.12030   0.12030
2 5.0   0.09083   0.09083
3 0.5  63.57390  63.55861
3 1.0  10.06008  10.06008
3 1.5   2.93785   2.93785
3 2.0   1.16206   1.16206
```

```
3  2.5   0.55148   0.55148
3  3.0   0.29562   0.29562
3  3.5   0.17291   0.17291
3  4.0   0.10800   0.10800
3  4.5   0.07100   0.07100
3  5.0   0.04864   0.04864
4  1.0  29.28820  29.28820
4  1.5   6.17284   6.17284
4  2.0   1.91294   1.91294
4  2.5   0.74689   0.74689
4  3.0   0.34026   0.34026
4  3.5   0.17311   0.17311
4  4.0   0.09570   0.09570
4  4.5   0.05644   0.05644
4  5.0   0.03506   0.03506
```

Both of these programs may be found in the applications\Laplace folder in the on-line archive.

Gamma and Psi Functions

The gamma and digamma (or psi) functions are defined in terms of integrals and also have known infinite series solutions as well as accurate approximations. The gamma function is defined by the integral:

$$\Gamma(z) = \int_0^\infty t^{z-1} e^{-t} dt \qquad (9.7)$$

The gamma function provides another test of improper integral evaluation using Gauss Quadrature. The same two functions are modified to handle this special case:

```
double GQ0Ia(double*A,double*W,int n,double
   f(double,double),double z)
  {
  int i;
  double q,t,x;
  for(q=i=0;i<n/2;i++)
    {
    x=(1.-A[i])/2.;
    t=1./x-1.;
    q+=W[i]*f(z,t)/x/x;
    x=(1.+A[i])/2.;
    t=1./x-1.;
    q+=W[i]*f(z,t)/x/x;
    }
  return(q/2.);
  }
double GQ0Ib(double*A,double*W,int n,double
   f(double,double),double z)
  {
```

```
  int i;
  double q,t,x;
  for(q=i=0;i<n/2;i++)
     {
     x=(1.-A[i])/2.;
     t=-log(x);
     q+=W[i]*f(z,t)/x;
     x=(1.+A[i])/2.;
     t=-log(x);
     q+=W[i]*f(z,t)/x;
     }
  return(q/2.);
  }
double f(double z,double t)
  {
  return(pow(t,z-1.)*exp(-t));
  }
```

The approximate solution is also simple:

```
double gamma(double x)
  {
  static double c[9]={1.,-0.577191652,0.988205891,-
   0.897056937,0.918206857,
    -0.756704078,0.482199394,-0.193527818,0.035868343};
  int i;
  double g,y,z;
  y=x;
  g=1.;
  while(y>2.)
     {
     y-=1.;
     g*=y;
     }
  while(y<1.)
     {
     g/=y;
     y+=1.;
     }
  z=c[8];
  for(i=7;i>=0;i--)
     {
     z*=y-1.;
     z+=c[i];
     }
  return(z*g);
  }
```

These constants may be found in Abramowitz and Stegun. The recurrence relationship is used to shift the value into the range 1 to 2, where this approximation is more than adequate. The quadrature results are quite accurate:

```
Gamma Function by Gauss Quadrature
 z    result1   result2    exact
0.5  1.772241  1.772241  1.772454
0.6  1.489167  1.489167  1.489193
0.7  1.298052  1.298052  1.298055
0.8  1.164229  1.164229  1.164230
0.9  1.068629  1.068629  1.068629
1.0  1.000000  1.000000  1.000000
1.1  0.951351  0.951351  0.951351
1.2  0.918169  0.918169  0.918169
1.3  0.897471  0.897471  0.897471
1.4  0.887264  0.887264  0.887264
1.5  0.886227  0.886227  0.886227
1.6  0.893515  0.893515  0.893516
1.7  0.908639  0.908639  0.908639
1.8  0.931384  0.931384  0.931384
1.9  0.961766  0.961766  0.961766
2.0  1.000000  1.000000  1.000000
2.1  1.046486  1.046486  1.046486
2.2  1.101802  1.101802  1.101803
2.3  1.166712  1.166712  1.166712
2.4  1.242169  1.242169  1.242169
2.5  1.329340  1.329340  1.329340
```

The digamma (or Psi) function is given by:

$$\psi(z) = \frac{d\ln[\Gamma(z)]}{dz} \tag{9.8}$$

This derivative is also equal to the following integral:

$$\psi(z) = -\gamma + \int_0^1 \left(\frac{1-t^{z-1}}{1-t}\right) dt \tag{9.9}$$

This integral (Equation 9.9) can also be calculated using Gauss Quadrature using the methods outlined in Chapter 3. The differential (Equation 9.8) can also easily be calculated using the gamma function listed above, as shown below:

```
double digamma(double z)
  {
  double dz;
  dz=0.0001;
  return((gamma(z+dz/2.)-gamma(z-dz/2.))/dz/gamma(z));
  }
```

The results are quite accurate:

```
digamma (Psi) function
 z    integra   differn
0.5  -1.96330  -1.96351
0.6  -1.54059  -1.54062
```

```
0.7  -1.22002  -1.22003
0.8  -0.96501  -0.96500
0.9  -0.75493  -0.75493
1.0  -0.57722  -0.57719
1.1  -0.42375  -0.42376
1.2  -0.28904  -0.28904
1.3  -0.16919  -0.16920
1.4  -0.06138  -0.06138
1.5   0.03649   0.03649
1.6   0.12605   0.12605
1.7   0.20855   0.20854
1.8   0.28499   0.28500
1.9   0.35618   0.35618
2.0   0.42278   0.42281
2.1   0.48534   0.48534
2.2   0.54429   0.54430
2.3   0.60004   0.60004
2.4   0.65290   0.65290
2.5   0.70316   0.70316
```

Both programs (gamma.c and digamma.c), along with a batch file to compile them, may be found in the applications\gamma folder.

Incomplete Gamma & Beta Functions

The incomplete gamma and beta functions appear in various derivations throughout applied mathematics. Although these can be calculated from infinite series, the coefficients are complicated and many terms are often required. These can also be estimated numerically by quadrature. The incomplete gamma function is defined by the following two integral:

$$\Gamma_x(a) = \int_0^x e^{-t} t^{a-1} dt \qquad (9.10)$$

The previous functions can be adapted as follows:

```
double GQ(double*A,double*W,int n,double
   f(double,double),double a,double x)
 {
 int i;
 double q,t;
 for(q=i=0;i<n/2;i++)
   {
   t=x*(1.-A[i])/2.;
   q+=W[i]*f(a,t);
   t=x*(1.+A[i])/2.;
   q+=W[i]*f(a,t);
   }
 return(x*q/2.);
 }
```

```
double f(double a,double t)
  {
  return(pow(t,a-1.)*exp(-t));
  }
double gstar(double a,double x)
  {
  return(pow(x,-a)*GQ(A4096,W4096,4096,f,a,x)/gamma(a));
  }
```

The output of the program (igamma.c) is listed below:

```
incomplete gamma function
  a   x    g*(a,x)
0.5 0.5 0.965349
0.5 1.0 0.842581
0.5 1.5 0.748391
0.5 2.0 0.674813
0.5 2.5 0.616305
0.5 3.0 0.568971
1.0 0.5 0.786939
1.0 1.0 0.632121
1.0 1.5 0.517913
1.0 2.0 0.432332
1.0 2.5 0.367166
1.0 3.0 0.316738
1.5 0.5 0.562144
1.5 1.0 0.427593
1.5 1.5 0.331157
1.5 2.0 0.261112
1.5 2.5 0.209521
1.5 3.0 0.170971
2.0 0.5 0.360816
2.0 1.0 0.264241
2.0 1.5 0.196522
2.0 2.0 0.148499
2.0 2.5 0.114032
2.0 3.0 0.088984
2.5 0.5 0.211760
2.5 1.0 0.150855
2.5 1.5 0.108871
2.5 2.0 0.079653
2.5 2.5 0.059109
2.5 3.0 0.044506
3.0 0.5 0.115101
3.0 1.0 0.080301
3.0 1.5 0.056638
3.0 2.0 0.040415
3.0 2.5 0.029196
3.0 3.0 0.021363
```

A map can be drawn using the graphics program mentioned previously:

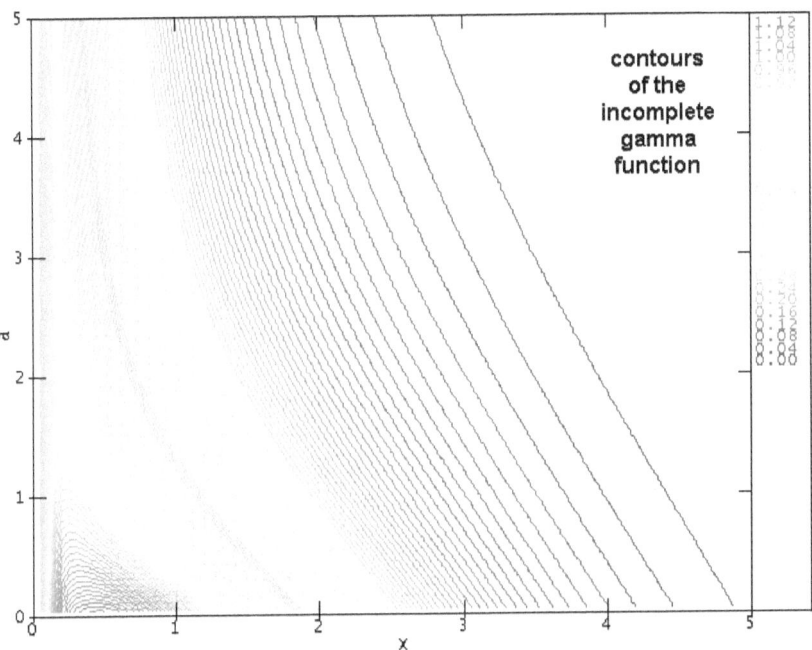

The incomplete beta function is defined by the following two integral:

$$B_x(a,b) = \int_0^x t^{a-1}(1-t)^{b-1} dt \qquad (9.11)$$

The same functions can be adapted as follows:

```
double GQ(double*A,double*W,int n,double
   f(double,double,double),double a,double b,double x)
  {
  int i;
  double q,t;
  for(q=i=0;i<n/2;i++)
    {
    t=x*(1.-A[i])/2.;
    q+=W[i]*f(a,b,t);
    t=x*(1.+A[i])/2.;
    q+=W[i]*f(a,b,t);
    }
  return(x*q/2.);
  }
double f(double a,double b,double t)
  {
  return(pow(t,a-1.)*pow(1.-t,b-1.));
  }
```

```
double ibeta(double a,double b,double x)
   {
   return(GQ(A4096,W4096,4096,f,a,b,x));
   }
```

The results are as follows:

```
incomplete beta function
 x    a    b   beta(a,b,x)
0.2  0.5  0.5  0.927200
0.2  0.5  1.0  0.894332
0.2  0.5  1.5  0.863553
0.2  0.5  2.0  0.834704
0.2  1.0  0.5  0.211146
0.2  1.0  1.0  0.200000
0.2  1.0  1.5  0.189639
0.2  1.0  2.0  0.180000
0.2  1.5  0.5  0.063648
0.2  1.5  1.0  0.059628
0.2  1.5  1.5  0.055912
0.2  1.5  2.0  0.052473
0.2  2.0  0.5  0.021507
0.2  2.0  1.0  0.020000
0.2  2.0  1.5  0.018612
0.2  2.0  2.0  0.017333
0.4  0.5  0.5  1.369304
0.4  0.5  1.0  1.264777
0.4  0.5  1.5  1.174483
0.4  0.5  2.0  1.096122
0.4  1.0  0.5  0.450807
0.4  1.0  1.0  0.400000
0.4  1.0  1.5  0.356828
0.4  1.0  2.0  0.320000
0.4  1.5  0.5  0.194821
0.4  1.5  1.0  0.168655
0.4  1.5  1.5  0.146685
0.4  1.5  2.0  0.128178
0.4  2.0  0.5  0.093979
0.4  2.0  1.0  0.080000
0.4  2.0  1.5  0.068370
0.4  2.0  2.0  0.058667
0.6  0.5  0.5  1.771990
0.6  0.5  1.0  1.549029
0.6  0.5  1.5  1.375810
0.6  0.5  2.0  1.239190
0.6  1.0  0.5  0.735089
0.6  1.0  1.0  0.600000
0.6  1.0  1.5  0.498012
```

TP2 can be used to create a 3D plot of these results.

Appendix A: Data Transformation Program

The following program manipulates orthogonal polynomials to arrive at coefficients that will smoothen, interpolate, extrapolate, integrate, and/or differentiate evenly-spaced data in a single step without resorting to matrix inversion. The polynomials P(n,x) have the property that:

$$\sum_{x=0}^{n} P_{nj}(x) P_{nk}(x) \neq 0 \quad j = k$$
$$\sum_{x=0}^{n} P_{nj}(x) P_{nk}(x) = 0 \quad j \neq k \tag{A.1}$$

Milne has shown that these polynomials are given by:[16]

$$P_{nm}(x) = \sum_{i=0}^{m} (-1)^i B(m,i) B(m+i,i) \frac{\Pi(x,i)}{\Pi(n,i)} \quad m = 0,1,2,\ldots,n \tag{A.2}$$

where $B(i,j)$ is the binomial coefficient and $\Pi(n,i)$ is the following product:

$$\Pi(n,i) = n(n-1)(n-2)\ldots(n-i+1) \tag{A.3}$$

The first 3 polynomials are listed below:

$$P_{no}(x) = 1$$
$$P_{n1}(x) = 1 - \frac{2x}{n} \tag{A.4}$$
$$P_{n2}(x) = 1 - \frac{6x}{n} + \frac{6x(x-1)}{n(n-1)}$$

The code is listed below and may be found in the examples\coefs folder in the on-line archive:

```
/* Determine coefficients that smoothen, interpolate,
     extrapolate, integrate
   and/or differentiate a set of equally-spaced points in
     one step.
   developed by Dudley J. Benton, Knoxville, Tennessee */

#define _CRT_SECURE_NO_DEPRECATE
#include <stdio.h>
#include <stdlib.h>
#include <malloc.h>
#include <float.h>
#define _USE_MATH_DEFINES
#include <math.h>
```

---

[16] Milne, W. E., *Numerical Analysis*, pp. 265-275 and 375-381, Princeton University Press, 1949.

```c
void iserror(int test,char*error,int
    line){if(test){fprintf(stderr,"error at line %i:
    %s\n",line,error);exit(1);}}
#define iferror(test) iserror(test,#test,__LINE__)

double binomial(int n,int m)
{ /* binomial coefficient */
 int i,j;
 double b=1.;
 if(n<m||m<0)
 return(0.);
 if(n==m||m<1)
 return(1.);
 if(m>n-m)
 m=n-m;
 for(i=1,j=n;i<=m&&i<j;i++,j--)
 {
 b*=j;
 b/=i;
 }
 return(b);
 }

double polynomial(int n,int m,double x)
 { /* m-th degree polynomial orthogonal over the finite
    set of n points */
 int i;
 double p=1.,q=1.;
 if(m!=0)
 {
 for(i=1;i<=m;i++)
 {
 x-=1;
 n--;
 if(n)
 q*=-x/n;
 p+=binomial(m,i)*binomial(m+i,i)*q;
 }
 }
 return(p);
 }

void derivative(double*C,int p,int d)
 { /* differentiate a polynomial */
 int i,j;
 for(j=1;j<=d;j++)
 {
 for(i=1;i<p;i++)
 C[i-1]=i*C[i];
```

```c
C[p-1]=0;
}
}
void Lagrange(double*A,int n)
 { /* expansion of the Largange polynomial */
 int i,j,k,l,new,old;
 double*C;
 C=calloc(n*2,sizeof(double));
 for(i=0;i<n;i++)
 {
 new=0;
 old=n;
 C[new]=1;
 for(j=1;j<n;j++)
 C[j+new]=0;
 for(j=l=0;j<n;j++)
 {
 if(j!=i)
 {
 l++;
 new=old;
 old=n-new;
 for(k=0;k<l;k++)
 C[k+new]=-C[k+old]*(j+1);
 C[l+new]=0;
 for(k=0;k<l;k++)
 C[k+1+new]+=C[k+old];
 for(k=0;k<=l;k++)
 C[k+new]/=i-j;
 }
 }
 for(j=0;j<n;j++)
 A[i*n+j]=C[j+new];
 }
 free(C);
 }

void Smooth(double*S,int p,int s)
 { /* smoothing coefficients for polynomial of degree p-
    1-s */
 int i,j,k,l,m;
 double*P,*Q,R;
 iferror(s<0);
 iferror(s>=p);
 memset(S,0,p*p*sizeof(double));
 if(s<1) /* if s=0 then [I] */
 {
 for(i=0;i<p;i++)
```

```
      S[p*i+i]=1;
      return;
      }
      P=calloc(p*p,sizeof(double));
      Q=calloc(p*p,sizeof(double));
      m=p-s;
      for(i=0;i<p;i++)
      for(j=0;j<m;j++)
      P[p*j+i]=Q[p*i+j]=polynomial(p,j,i+1.);
      for(i=0;i<m;i++)
      {
      for(R=j=0;j<p;j++)
      R+=P[p*i+j]*P[p*i+j];
      for(j=0;j<p;j++)
      P[p*i+j]/=R;
      }
      for(i=l=0;i<p;i++)
      for(j=0;j<p;j++,l++)
      for(k=0;k<m;k++)
      S[l]+=Q[p*i+k]*P[p*k+j];
      free(P);
      free(Q);
      }

    void Coefficients(double*C,int p,int d,int s,double i)
      {
      int j,k,l;
      double*A,*B,*D,*S;
      iferror(p<1);
      iferror(p>20);
      iferror(d<-1);
      iferror(s<0);
      iferror(s>=p);
      iferror(d>0&&s+d>=p);
      memset(C,0,p*sizeof(double));
      A=calloc(p*p,sizeof(double));
      B=calloc(p,sizeof(double));
      D=calloc(p,sizeof(double));
      S=calloc(p*p,sizeof(double));
      Lagrange(A,p);
      if(d<0)
      {
      B[0]=i;
      for(k=1;k<p;k++)
      B[k]=B[k-1]*i;
      for(k=0;k<p;k++)
      B[k]=(B[k]-1)/(k+1);
      for(k=0;k<p;k++)
      for(j=0;j<p;j++)
```

```
D[k]+=A[p*k+j]*B[j];
if(fabs(i-1.)>FLT_EPSILON)
for(k=0;k<p;k++)
D[k]/=fabs(i-1.);
}
else
{
B[0]=1;
for(k=1;k<p;k++)
B[k]=B[k-1]*i;
if(d>0)
for(k=0;k<p;k++)
derivative(A+p*k,p,d);
for(k=l=0;k<p;k++)
for(j=0;j<p;j++,l++)
D[k]+=A[l]*B[j];
}
if(s>0)
{
Smooth(S,p,s);
for(k=l=0;k<p;k++)
for(j=0;j<p;j++,l++)
C[k]+=S[l]*D[j];
}
else
memcpy(C,D,p*sizeof(double));
free(A);
free(B);
free(D);
free(S);
}
int main(int argc,char**argv,char**envp)
{
int d,j,p,s;
double C[20],i;
printf("transformation coefficients for equally-spaced
   data\n");
printf("p=points, i=point, d=differentiate (-1 for
   integrate),s=smoothing\n");
if(argc!=5)
{
printf("usage: coefs p i d s as in\ncoefs 6 7 0 1\n");
return(1);
}
p=atoi(argv[1]);
i=atof(argv[2]);
d=atoi(argv[3]);
s=atoi(argv[4]);
```

```c
printf("p=%i,i=%lG,d=%i,s=%i\n",p,i,d,s);
iferror(p<2);
iferror(p>20);
iferror(d<-1);
iferror(d>=p);
iferror(s<0);
iferror(s>=p);
iferror(d>0&&s+d>=p);
iferror(i<-p);
iferror(i>2*p);
Coefficients(C,p,d,s,i);
printf("%lG",C[0]);
for(j=1;j<p;j++)
printf(",%lG",C[j]);
printf("\n");
return(0);
}
```

Appendix B. Newton-Cotes Coefficient Program

The following program calculates the coefficients for any order of Newton-Cotes integration without inverting a matrix. It is a subset of the program presented in Appendix A. The files may be found in the examples\NewtonCotes folder of the on-line archive.

```c
/* Coefficients for Newton-Cotes Integration
 developed by Dudley J. Benton, Knoxville, Tennessee */

#define _CRT_SECURE_NO_DEPRECATE
#include <stdio.h>
#include <stdlib.h>
#include <malloc.h>
#include <float.h>
#define _USE_MATH_DEFINES
#include <math.h>

void iserror(int test,char*error,int
    line){if(test){fprintf(stderr,"error at line %i:
    %s\n",line,error);exit(1);}}
#define iferror(test) iserror(test,#test,__LINE__)

void Lagrange(double*A,int n)
 { /* expansion of the Largange polynomial */
 int i,j,k,l,new,old;
 double*C;
 C=calloc(n*2,sizeof(double));
 for(i=0;i<n;i++)
 {
 new=0;
 old=n;
 C[new]=1;
 for(j=1;j<n;j++)
 C[j+new]=0;
 for(j=l=0;j<n;j++)
 {
 if(j!=i)
 {
 l++;
 new=old;
 old=n-new;
 for(k=0;k<l;k++)
 C[k+new]=-C[k+old]*(j+1);
 C[l+new]=0;
 for(k=0;k<l;k++)
 C[k+1+new]+=C[k+old];
 for(k=0;k<=l;k++)
 C[k+new]/=i-j;
 }
```

```c
 }
 for(j=0;j<n;j++)
 A[i*n+j]=C[j+new];
 }
 free(C);
 }

void Coefficients(double*C,int p)
 {
 int j,k;
 double*A,*B;
 iferror(p<1);
 memset(C,0,p*sizeof(double));
 A=calloc(p*p,sizeof(double));
 B=calloc(p,sizeof(double));
 Lagrange(A,p);
 B[0]=p;
 for(k=1;k<p;k++)
 B[k]=B[k-1]*p;
 for(k=0;k<p;k++)
 B[k]=(B[k]-1)/(k+1);
 for(k=0;k<p;k++)
 for(j=0;j<p;j++)
 C[k]+=A[p*k+j]*B[j];
 for(k=0;k<p;k++)
 C[k]/=p-1;
 free(A);
 free(B);
 }

int main(int argc,char**argv,char**envp)
 {
 int j,p;
 double*C;
 printf("Newton-Cotes integration coefficients\n");
 if(argc!=2)
 {
 printf("usage: NewtonCotes 6\n");
 return(1);
 }
 p=atoi(argv[1]);
 iferror(p<2);
 C=calloc(p,sizeof(double));
 Coefficients(C,p);
 printf("%lG",C[0]);
 for(j=1;j<p;j++)
 printf(",%lG",C[j]);
 printf("\n");
 return(0);
```

}

The coefficients for orders up to 16 are listed below. The coefficients up to 60th may be found in the examples\NewtonCotes folder in the on-line archive.

```
double NC2[]={1./2.,1./2.};
double NC3[]={1./6.,2./3.,1./6.};
double NC4[]={1./8.,3./8.,3./8.,1./8.};
double NC5[]={7./90.,32./90.,12./90.,32./90.,7./90.};
double NC6[]={19./288.,75./288.,50./288.,50./288.,
    75./288.,19./288.};
double NC7[]={41./840.,216./840.,27./840.,272./840.,
    27./840.,216./840.,41./840.};
double NC8[]={751./17280.,3577./17280.,1323./17280.,
    2989./17280.,2989./17280.,1323./17280.,3577./17280.,
    751./17280.};
double NC9[]={989./28350.,5888./28350.,-928./28350.,
    10496./28350.,-4540./28350.,10496./28350.,
    -928./28350.,5888./28350.,989./28350.};
double NC10[]={2857./89600.,15741./89600.,1080./89600.,
    19344./89600.,5778./89600.,5778./89600.,
    19344./89600.,1080./89600.,15741./89600.,
    2857./89600.};
double NC11[]={16067./598752.,106300./598752.,
    -48525./598752.,272400./598752.,
    -260550./598752.,427368./598752.,
    -260550./598752.,272400./598752.,
    -48525./598752.,106300./598752.,16067./598752.};
double NC12[]={0.0249332309119641,0.1548553585207226,
    -0.0371692317937990,0.2896582547949741,
    -0.1101780891754860,0.1779004767416227,
    0.1779004767416230,-0.1101780891754863,
    0.2896582547949741,-0.0371692317937991,
    0.1548553585207226,0.0249332309119641};
double NC16[]={0.0170872997716264,0.1285073786774053,
    -0.1127229050595076,0.5070427082102271,
    -0.7562931148441474,1.1913603495068070,
    -0.9680052114962560,0.4930234952338425,
    0.4930234952338438,-0.9680052114962570,
    1.1913603495068090,-0.7562931148441480,
    0.5070427082102273,-0.1127229050595076,
    0.1285073786774054,0.0170872997716265};
```

Appendix C. Gauss Quadrature Weights & Abscissas

As discussed in Chapter 3, the abscissas for Gauss Quadrature (GQ) are the roots of the n+1 degree Legendre Polynomial and the weights are values of the nth degree polynomial at each of the roots. The abscissas are found by locating the roots and the weights are found by plugging these values into the appropriate polynomial. The following program accomplishes this:

```
/* Find coefficients (weights and abscissas) for Gauss
   Quadrature
   This version uses extended-precision floating-point
   operations.
   developed by Dudley J. Benton, Knoxville, Tennessee
   */
#define _CRT_SECURE_NO_DEPRECATE
#include <stdio.h>
#include <stdlib.h>
#include <string.h>
#include <malloc.h>
#include <float.h>
#define _USE_MATH_DEFINES
#include <math.h>
#include "xreal.hpp"
#include "real.cpp"

real tiny;

void LegendrePolynomial(int n,real X,real*P3,real*Q3)
  {
  int i;
  real P1,P2,Q1,Q2;
  *P3=1.;
  *Q3=0.;
  if(n<1)
    return;
  P2=*P3;
  Q2=*Q3;
  *P3=X;
  *Q3=1.;
  if(n<2)
    return;
  for(i=2;i<=n;i++)
    {
    P1=P2;
    Q1=Q2;
    P2=*P3;
    Q2=*Q3;
    *P3=2.*X*P2-P1-(X*P2-P1)/i;
    *Q3=Q1+(2*i-1)*P2;
    }
```

```c
    }

real RefineRoot(int n,real X1,real X2)
  {
  int i;
  real dP,P,Xo;
  static real X;
  X=(X1+X2)/2.;
  for(i=0;i<128;i++)
     {
     LegendrePolynomial(n,X,&P,&dP);
     if(fabs(dP)<0.5)
       break;
     Xo=X;
     X=fmax(X1,fmin(X2,X-P/dP));
     if(fabs(P)<=tiny||fabs(X-Xo)<=tiny)
       return(X);
     }
  X=2.;
  return(X);
  }

int InsertRoot(real*Roots,int nr,real Root)
  {
  int i,j;
  for(i=0;i<nr;i++)
    if(fabs(Roots[i]-Root)<tiny)
      return(0);
  for(i=1;i<nr;i++)
    if(Roots[i-1]<Root&&Root<Roots[i])
      break;
  for(j=nr;j>i;j--)
    Roots[j]=Roots[j-1];
  Roots[i]=Root;
  return(i);
  }

int FindRoots(int n,real*Roots,real*Weights)
  {
  int i,j,k,nr;
  real dP,dX,P,Root,X,X1,X2;
  printf("computing coefficients\n");
  if(n<2)
    return(0);
  nr=1;
  if(n%2)
    Roots[0]=0;
  else
    {
```

```
    X1=0.;
    X2=1.3/pow(n,0.94);
    Roots[0]=RefineRoot(n,X1,X2);
    if(n<3)
      goto weights;
    }
  X1=1.-2.1/pow(n,1.9);
  X2=1.;
  Roots[nr++]=RefineRoot(n,X1,X2);
  if(n<5)
    goto weights;
  printf("\r%i roots found out of %i",nr,(n+1)/2);
  while(nr<(n+1)/2)
    {
    for(i=1;i<nr;i++)
      {
      k=(n+1)/2-nr+1;
      dX=(Roots[i]-Roots[i-1])/k;
      X2=Roots[i-1];
      for(j=0;j<k;j++)
        {
        X1=X2;
        X2=X2+dX;
        Root=RefineRoot(n,X1,X2);
        if(Root>1.5)
          continue;
        if(!InsertRoot(Roots,nr,Root))
          continue;
        nr++;
        printf("\r%i roots found out of %i",nr,(n+1)/2);
        break;
        }
      if(nr>=(n+1)/2)
        goto weights;
      }
    }
  weights:
  if(n>4)
    printf("\n");
  for(i=0;i<(n+1)/2;i++)
    {
    X=Roots[i];
    LegendrePolynomial(n,X,&P,&dP);
    Weights[i]=2/(1-X*X)/dP/dP;
    }
  return((n+1)/2);
  }

int main(int argc,char**argv,char**envp)
```

```
{
int i,n;
real*A,*W;
printf("Gauss Quadrature Weights and Abscissas\n");
if(argc!=2)
   {
   printf("usage: GQUAD 123\n");
   return(1);
   }
n=atoi(argv[1]);
if(n<2)
   {
   printf("n must be at least 2\n");
   return(1);
   }
A=(real*)calloc(((n+1)/2),sizeof(real));
W=(real*)calloc(((n+1)/2),sizeof(real));
tiny=DBL_EPSILON*DBL_EPSILON;
FindRoots(n,A,W);
printf("abscissas\n");
for(i=0;i<(n+1)/2;i++)
   printf("%s\n",xtoa(&A[i].x));
printf("weights\n");
for(i=0;i<(n+1)/2;i++)
   printf("%s\n",xtoa(&W[i].x));
return(0);
}
```

The roots up to 4096 may be found in the examples\GaussQuadrature folder in the on-line archive. The weights and abscissas for orders 3 through 6 were listed in Chapter 3. The weights and abscissas for orders 7 through 12 are listed below:

```
double A7[]={0.000000000000,0.405845151377,
   0.741531185599,0.949107912343};
double W7[]={0.417959183673,0.381830050505,
   0.279705391489,0.129484966169};
double A8[]={0.183434642496,0.525532409916,
   0.796666477414,0.960289856498};
double W8[]={0.362683783378,0.313706645878,
   0.222381034453,0.101228536290};
double A9[]={0.000000000000,0.324253423404,
   0.613371432701,0.836031107327,0.968160239508};
double W9[]={0.330239355001,0.312347077040,
   0.260610696403,0.180648160695,0.081274388362};
double A10[]={0.148874338982,0.433395394129,
   0.679409568299,0.865063366689,0.973906528517};
double W10[]={0.295524224715,0.269266719310,
   0.219086362516,0.149451349151,0.066671344309};
double A12[]={0.125233408511,0.367831498998,
```

```
     0.587317954287,0.769902674194,0.904117256370,
     0.981560634247};
double W12[]={0.249147045813,0.233492536538,
     0.203167426723,0.160078328543,0.106939325995,
     0.047175336387};
```

The odd orders are not as efficient as the even ones, which is why you rarely see the high order odd coefficients.

Static vs. Global Data in C

Programming quadrature often requires large data statements, something you can't do in VBA® and is really convoluted in VB.Net®. Data statements are natural structures in all other programming languages, which is yet another reason to graduate from BASIC to a professional language.[17] In C, all variables with global scope (i.e., outside a function) are inherently static. In spite of this fact, you often see code and/or data with the qualifier "static" at the global level. This indicates that whoever wrote the code doesn't understand C or how compilers work, as this qualifier is ignored in this context.

At the local level (i.e., within the context of a function), all variables are *discardable* unless qualified as *static*. All discardable objects are given temporary locations on the stack.[18] This means that in the code below:

```
function MyQuad(double f(double x),double a,double b)
{
int i;
double q;
double A[4]={0.,0.4,0.6,1.0};
double W[4]={0.25,0.25,0.25,0.25};
for(q=i=0;i<4;i++)
  q+=W[i]*f(a+A[i]*(b-a));
return(q);
}
```

the statements A[4]={... and W[4]={... will be *executed* each and every time you enter the function! If you look at the assembler (i.e., machine language instructions) that the compiler generates, you will see the constants stored somewhere in global memory are copied into the temporary locations on the stack every time. This may take as long as the function itself. If you mean for

---

[17] Never forget that BASIC stands for *Beginner's* All-Purpose Symbolic Instruction Code. It's the programmer's equivalent of Pull-Ups®, the popular transition from diapers to potty training.

[18] Intel® processors are *stack* machines. Most modern processors are. The *stack* is a designated block of memory that facilitates calling and returning from procedures, passing parameters (i.e., arguments), and temporary storage. It's called a *stack* because it's most often accessed like cafeteria trays: last on is first off, first on is last off.

these constants to always be available and have these same values within the context of a function, then qualify them with static, as in:
```
static double A[4]={0.,0.4,0.6,1.0};
static double W[4]={0.25,0.25,0.25,0.25};
```
Remember... you only need to do this at the local level.

Extended Precision

You may notice that some of the weights and abscissas contained in the examples exceed that which can be produced with traditional code. That's because I ran those out specifically for this purpose using extended precision. The Microsoft® C compilers of the past decade no longer support 80-bit floating-point numbers, but Intel processors always have and still do. I still have a copy of the old Microsoft® C compiler that does. Of course, it produces 16-bit executables that won't run on a 64-bit O/S except in a virtual box. Even 80-bit only gets you so far.

There is another way... extended precision. Several sources of extended precision math code may be found on the Web. The ones I have been able to find are only compatible with the GNU compiler, gcc, which less common than three-headed talking chickens and far less useful. Pass one of these codes to the Microsoft® C compiler and your machine will self-destruct in a cry of agony.

Fortunately for you, I have developed my own extended precision math code, which you are welcome to have and use. Here's how it works...

```
/* define extended precision real functions
   number  digits
     of      of
   bytes  precision
     8      15  <- minimum
    12      25
    16      35    note: set the number of bytes for
    20      44          the xreal in the typedef below
    24      54
    28      63    note: the number of bytes is not
    32      73          limited to multiples of 4
    36      82
    40      92
    44     102
    48     111
    52     121
    56     131
    60     140
    64     150  */
typedef struct xreal{BYTE b[24];}xreal;
```

You will find all of these files in the xreal folder. This definition is in file xreal.hpp. IEEE 754 double precision is 64-bit or 8 bytes. As shown in the preceding table, this yields approximately 15 significant digits. If you dimension b[8] in the xreal structure, the code will just use the FPU and be a whole lot faster. If you dimension b[12], b[16], etc., extended precision will take over.

The first 8 bytes of xreal are exactly the same as IEEE 754 or Intel® 64-bit double precision floating-point numbers. The bytes beyond 8 are additional mantissa. The exponent part is identical for xreal of any size. This is different from the Intel® 80-bit numbers that have exponents of 2^{-1024} to 2^{+1023} or about 10^{-308} to 10^{+308}. The header file xreal.hpp defines a variety of functions, as listed below. The code to implement these calculations is in xreal.cpp.

char*	xtoa(xreal*)	encode an xreal
char*	xtohex(xreal*x)	xreal to hex
int	intx(xreal*)	convert xreal to int
int	nintx(xreal*)	convert xreal to nearest int
int	xalmost(xreal*x,xreal*y)	xreals almost the same?
int	xclose(xreal*x,xreal*y)	xreals close?
int	xcmp(xreal*,xreal*)	compare 2 xreals
int	xcmpz(xreal*x)	compare xreal to zero
int	xisinf(xreal*x)	is?inf
int	xisnan(xreal*x)	is?nan
int	xiszero(xreal*)	test for not zero
int	xnearly(xreal*x,xreal*y)	xreals nearly the same?
int	xsame(xreal*,xreal*)	compare 2 xreals
double	xtod(xreal*)	convert xreal to double
xreal	atox(char*)	decoded an xreal
xreal	dtox(double d)	convert double to xreal
xreal	itox(int i)	convert int to xreal
xreal	xabs(xreal*)	absolute value
xreal	xacos(xreal*)	arccos(x)
xreal	xadd(xreal*,xreal*)	add 2 xreals
xreal	xasin(xreal*)	arcsin(x)
xreal	xatan(xreal*)	arctan(x)
xreal	xatan2(xreal*,xreal*)	arctan(x/y)
xreal	xchs(xreal*)	change signs
xreal	xcos(xreal*)	cos(x)
xreal	xdiv(xreal*,xreal*)	divide 2 xreals
xreal	xdot(xreal*p,xreal*q,int n)	d=sum(p[i]*q[i]),i=1..n
xreal	xe()	return exp(1)
xreal	xexp(xreal*)	return exp(x)
xreal	xhalf(xreal*x)	divide an xreal by 2
xreal	xinf()	inf
xreal	xlarge()	largest possible xreal

xreal	xlog(xreal*)	return ln(x)
xreal	xlog10(xreal*)	return log(x)
xreal	xmax(xreal*x,xreal*y)	max(x,y)
xreal	xmin(xreal*x,xreal*y)	min(x,y)
xreal	xmul(xreal*,xreal*)	multiply 2 xreals
xreal	xnan()	nan
xreal	xone()	return 1
xreal	xpi()	return π
xreal	xpow(xreal*,xreal*)	power x^y
xreal	xsin(xreal*)	sin(x)
xreal	xsmall()	smallest possible xreal >0
xreal	xsqrt(xreal*)	square root of xreal
xreal	xsub(xreal*,xreal*)	subtract 2 xreals
xreal	xtan(xreal*)	tan(x)
xreal	xten()	return 10
xreal	xtoi(xreal*,int)	raise xreal to int
xreal	xtwice(xreal*x)	multiply an xreal by 2
xreal	xtwo()	return 2
xreal	xzero()	zero
void	xset(xreal*p,int n,xreal*q)	set
void	xcopy(xreal*p,xreal*q,int n)	copy

The following code adds two extended floating-point numbers:

```
xreal xadd(xreal*x,xreal*y) /* add two xreals */
  {
  BYTE mask;
  int i,j,k,xe,xs,ye,ys,ze,zs;
  xreal xm,ym,zm;
  static xreal s;
  union{WORD w;BYTE b[2];}exponent,u;

  if(sizeof(xreal)==8)
     {
     xd.d=*(double*)x;
     xd.d+=*(double*)y;
     return(xd.x);
     }
  if(xiszero(y)) /* if y=0 then return x */
     return(*x);
  if(xiszero(x)) /* if x=0 then return y */
     return(*y);

  s=*x;
  exponent.b[1]=(s.b[sizeof(xreal)-1]>>4)&0x07; /*
    unpack xreal */
```

```
exponent.b[0]=(s.b[sizeof(xreal)-
 1]<<4)|(s.b[sizeof(xreal)-2]>>4);
xe=exponent.w-1023-(sizeof(xreal)*8-11);
xm=xzero();
for(i=0;i<sizeof(xreal)-1;i++)
  xm.b[i]=s.b[i];
xm.b[i-1]&=0x0F;
xm.b[i-1]|=0x10;
xs=xisneg(&s);

exponent.b[1]=(y->b[sizeof(xreal)-1]>>4)&0x07; /*
 unpack xreal */
exponent.b[0]=(y->b[sizeof(xreal)-1]<<4)|(y-
 >b[sizeof(xreal)-2]>>4);
ye=exponent.w-1023-(sizeof(xreal)*8-11);
ym=xzero();
for(i=0;i<sizeof(xreal)-1;i++)
  ym.b[i]=y->b[i];
ym.b[i-1]&=0x0F;
ym.b[i-1]|=0x10;
ys=xisneg(y);

if(ye>xe) /* make x the one with the larger exponent
 */
  {
  zs=xs;
  ze=xe;
  zm=xm;
  xs=ys;
  xe=ye;
  xm=ym;
  ys=zs;
  ye=ze;
  ym=zm;
  s=*y;
  }

if(xe-ye>sizeof(xreal)*8-11) /* If the exponents
 differ by more than the */
  return(s); /* number of significant digits don't
 bother adding them. */

if(xs) /* if x<0 fix signed integer */
  {
  for(i=0;i<sizeof(xreal);i++)
    xm.b[i]^=0xFF; /* The negative of a number is
 found by */
  u.b[1]=0; /* XORing it with -1 and adding 1. */
  u.b[0]=xm.b[0];
```

```c
   u.w+=1;
   xm.b[0]=u.b[0];
   for(i=1;i<sizeof(xreal);i++) /* add with carry */
      {
      u.b[0]=u.b[1];
      u.b[1]=0;
      if(u.w==0)
         break;
      u.w+=xm.b[i];
      xm.b[i]=u.b[0];
      }
   }

if(ys) /* if y<0 fix signed integer */
   {
   for(i=0;i<sizeof(xreal);i++)
      ym.b[i]^=0xFF; /* The negative of a number is found by */
   u.b[1]=0; /* XORing it with -1 and adding 1. */
   u.b[0]=ym.b[0];
   u.w+=1;
   ym.b[0]=u.b[0];
   for(i=1;i<sizeof(xreal);i++) /* add with carry */
      {
      u.b[0]=u.b[1];
      u.b[1]=0;
      if(u.w==0)
         break;
      u.w+=ym.b[i];
      ym.b[i]=u.b[0];
      }
   }

while(ye<xe) /* shift y to the right until the two exponents are equal */
   {
   if(xe-ye>=8) /* shift one BYTE at a time */
      {
      for(i=0;i<sizeof(xreal)-1;i++)
         ym.b[i]=ym.b[i+1];
      if(ys)
         ym.b[i]=0xFF;
      ye+=8;
      }
   else /* shift one bit at a time */
      {
      for(i=0;i<sizeof(xreal)-1;i++)
         {
         ym.b[i]>>=1;
```

```
      if(ym.b[i+1]&1)
         ym.b[i]|=0x80;
      }
    ym.b[i]>>=1;
    if(ys)
      ym.b[i]|=0x80;
    ye++;
    }
  }

u.b[1]=0; /* add the two integers */
for(i=0;i<sizeof(xreal);i++)
  {
  u.b[0]=u.b[1];
  u.b[1]=0;
  u.w+=xm.b[i];
  u.w+=ym.b[i];
  xm.b[i]=u.b[0];
  }

if(xisneg(&xm)) /* convert to positive integer */
  {
  xs=1;
  for(i=0;i<sizeof(xreal);i++)
    xm.b[i]^=0xFF;
  u.b[1]=0;
  u.b[0]=xm.b[0];
  u.w+=1;
  xm.b[0]=u.b[0];
  for(i=1;i<sizeof(xreal);i++)
    {
    u.b[0]=u.b[1];
    u.b[1]=0;
    if(u.w==0)
      break;
    u.w+=xm.b[i];
    xm.b[i]=u.b[0];
    }
  }
else
  xs=0;

k=-1; /* determine position of first non-zero bit of
 result */
j=8*sizeof(xreal);
for(i=sizeof(xreal)-1;i>=0&&k<0;i--)
  {
  for(mask=0x80;mask;mask>>=1,j--)
    {
```

```
      if(xm.b[i]&mask)
        {
        k=j;
        break;
        }
      }
    }
  if(k<0) /* check for zero result (underflow) */
    {
    s=xzero();
    return(s);
    }
  while(k>sizeof(xreal)*8-11) /* shift result right
    until normalized */
    {
    if(k-(sizeof(xreal)*8-11)>=8) /* shift one BYTE at a
    time */
      {
      for(i=0;i<sizeof(xreal)-1;i++)
        xm.b[i]=xm.b[i+1];
      xm.b[i]=0x00;
      xe+=8;
      k-=8;
      }
    else /* shift one bit at a time */
      {
      for(i=0;i<sizeof(xreal)-1;i++)
        {
        xm.b[i]>>=1;
        if(xm.b[i+1]&0x01)
          xm.b[i]|=0x80;
        }
      xm.b[i]>>=1;
      xe++;
      k--;
      }
    }
  while(k<sizeof(xreal)*8-11) /* shift result left until
    normalized */
    {
    if((sizeof(xreal)*8-11)-k>=8) /* shift one BYTE at a
    time */
      {
      for(i=sizeof(xreal)-1;i>0;i--)
        xm.b[i]=xm.b[i-1];
      xm.b[i]=0;
```

```
      xe-=8;
      k+=8;
      }
    else /* shift one bit at a time */
      {
      for(i=sizeof(xreal)-1;i>0;i--)
        {
        xm.b[i]<<=1;
        if(xm.b[i-1]&0x80)
          xm.b[i]|=1;
        }
      xm.b[i]<<=1;
      xe--;
      k++;
      }
    }
  xm.b[sizeof(xreal)-2]&=0x0F; /* copy result to x */
  exponent.w=(WORD)(xe+1023+(sizeof(xreal)*8-11));
  xm.b[sizeof(xreal)-2]|=(exponent.b[0]<<4);
  xm.b[sizeof(xreal)-
    1]=(exponent.b[1]<<4)|(exponent.b[0]>>4);
  s=xm;
  if(xs) /* set sign bit */
    s.b[sizeof(xreal)-1]|=0x80;

  return(s);
  }
```

You will find no copyright notice in these files other than my name, because the operations are not based on the work of anyone else. I first developed these operations in assembler to run on the HP-1000 minicomputer at the TVA Engineering Laboratory in Norris around 1982. I published them in 1992 after translating them into Intel® assembler.

While you can use these directly, you don't have to. I have also provided a C++ wrapper in the form of type real. The xreal is inside the class real. The class real facilitates access to the xreal operations that can be used just like common double precision ones. There are also some predefined constants, including: pi, e, gamma, golden, and ln(10).

The real class is as follows, which shows the supported operators:

```
class real
  {
  public:
  inline real d2r(double d){x=dtox(d);return(*this);}
  real   operator = (double d){return d2r(d);};
  friend real fabs(real);
  friend real fchs(real);
  friend real sqrt(real);
```

```
        friend real exp(real);
        friend real log(real);
        friend real pow(real,real);
        friend int operator  < (real,real);
        friend int operator  < (real,double);
        friend int operator  > (real,real);
        friend int operator  > (real,double);
        friend int operator != (real,real);
        friend int operator <= (real,real);
        friend int operator == (real,real);
        friend int operator >= (real,real);
        friend real operator * (real,real);
        friend real operator * (real,double);
        friend real operator * (double,real);
        friend real operator * (int,real);
        friend real operator * (real,int);
        friend real operator + (real,real);
        friend real operator + (real,double);
        friend real operator + (double,real);
        friend real operator - (real,real);
        friend real operator - (real,double);
        friend real operator - (double,real);
        friend real operator / (real,real);
        friend real operator / (real,double);
        friend real operator / (double,real);
        friend real operator / (real,int);
        xreal x;
     };
```

The results for 20-pt Gauss Quadrature are:

```
gquad 20
Gauss Quadrature Weights and Abscissas
computing coefficients
10 roots found out of 10
abscissas
+7.6526521133497333754640409398838211004796266813497499827E-002
+2.2778585114164507808049619536857462474308893768292747007E-001
+3.7370608871541956067254817702492723739574632170568270822E-001
+5.1086700195082709800436405095525099842549132920242682777E-001
+6.3605368072651502545283669622628593674338911679936845644E-001
+7.4633190646015079261430507035564159031073067956917643545E-001
+8.3911697182221882339452906170152068532962936506563736371E-001
+9.1223442825132590586775244120329811304918479742369176473E-001
+9.6397192727791379126766613119727722191206032780618884505E-001
+9.9312859918509492478612238847132027822264713090165588633E-001
weights
+1.5275338713072585069808433195509759349194864511237859526E-001
+1.4917298647260374678782873700196943669267990408136831628E-001
+1.4209610931838205132929832506716493303451541339202030150E-001
+1.3168863844917662689849449974816313491611051114698352684E-001
+1.181945319615184173123773771138228700504121954896877299E-001
+1.0193011981724043503675013548034987616669165602339255654E-001
```

```
+8.32767415767047487247581432220462061001778285831632892 41E-002
+6.26720483341090635695065351870416063516010765784363642 11E-002
+4.06014298003869413310399522749321098790906399899515355 39E-002
+1.76140071391521183118619623518528163621431055433367333 24E-002
```

Orders up to 4096 may be found in the \examples\GaussQuadrature folder.

Appendix D. Cooling Tower Demand

As mentioned in Chapters 3 and 9, cooling tower demand is calculated using numerical integration. The following program not only illustrates 3 different calculation methods (4-pt Chebyshev, 10-pt Lobatto, 4th order Runge-Kutta), but also includes a variety of moist air property function required for these calculations:

```
/* The purpose of this program is to compute KaV/L using
   3 methods
   This program was developed by Dudley J. Benton */

#define _CRT_SECURE_NO_DEPRECATE
#include <stdio.h>
#include <stdlib.h>
#include <float.h>
#define _USE_MATH_DEFINES
#include <math.h>

/* thermodynamic constants */

double Patm =14.696  ; /* change this and the properties
   will change accordingly */
double Cpa  =0.2406  ; /* constant pressure specific
   heat of dry air */
double Cpg  =0.427933; /* temperature variation of water
   vapor enthalpy */
double Cpw  =1.0;      ; /* constant pressure specific
   heat of water */
double Two  =32.018  ; /* triple point of water */
double Hwo  =1061.39 ; /* enthalpy of water vapor at 0øF
   */
double Rmw  =0.62198 ; /* ratio of water to air
   molecular weights */
double Lewis=1.2       ; /* Lewis Number */

/* moist air properties based on Hyland & Wexler */

double fs(double T,double P)
  {
  return(((((-4.55447E-10*T+9.400757E-09*P+1.282159E-
    07)*T-1.762686E-06*P+6.35199E-06)*T+3.18886E-
    04*P+1.000104);
  }

double fPwsat(double T)
  {
  double P1=-3.99087,P2=1.81071E-1,P3=-3.63953E-
    3,P4=2.7966E-5,
```

```c
        P5=-6.73798E-8,P6=9.27329E-11,Q1=-3.22286E-
    2,Q2=4.61283E-4;

    return(exp((((((P6*T+P5)*T+P4)*T+P3)*T+P2)*T+P1)/((Q2
    *T+Q1)*T+1.))*fs(T,Patm));
    }

double fTwsat(double P)
    {
    double
      P1=101.729,P2=95.4163,P3=33.8827,P4=5.24155,P5=0.3474
      41,

      P6=0.0167944,P7=0.000362087,Q1=0.609043,Q2=0.110268,T
      ;
    P=log(P);

      T=((((((P7*P+P6)*P+P5)*P+P4)*P+P3)*P+P2)*P+P1)/((Q2*P
      +Q1)*P+1.);
    P-=log(fs(T,Patm));

      T=((((((P7*P+P6)*P+P5)*P+P4)*P+P3)*P+P2)*P+P1)/((Q2*P
      +Q1)*P+1.);
    return(T);
    }

double fWsat(double T)
    {
    double P;
    P=fPwsat(T);
    return(Rmw*P/(Patm-P));
    }

double fPvw(double W) /* partial pressure of water vapor
    from humidity ratio */
    {
    return(W*Patm/(Rmw+W));
    }

double fHdbw(double Tdb,double W)
    {
    return(Cpa*Tdb+W*(Hwo+Cpg*Tdb));
    }

double fHtwb(double Twb)
    {
    return(fHdbw(Twb,fWsat(Twb)));
    }
```

```
double Ftwbh(double H)
  {
  int i;
  double Twb,Twb1,Twb2;
  Twb1=-30.;
  Twb2=fTwsat(Patm)-1.;
  for(i=0;i<25;i++)
    {
    Twb=(Twb1+Twb2)/2.;
    if(fHtwb(Twb)<H)
      Twb1=Twb;
    else
      Twb2=Twb;
    }
  return(Twb);
  }

double fWdwb(double Tdb,double Twb)
  {
  double X,Y;
  if(Tdb<=Twb)
    return(fWsat(Tdb));
  X=(Hwo+Cpw*(Two-Twb)+Cpg*Twb)*fWsat(Twb)-Cpa*(Tdb-
    Twb);
  Y=Hwo+Cpw*(Two-Twb)+Cpg*Tdb;
  return(X/Y);
  }

double fPvdwb(double Tdb,double Twb)
  {
  return(fPvw(fWdwb(Tdb,Twb)));
  }

double fHdwb(double Tdb,double Twb)
  {
  return(fHdbw(Tdb,fWdwb(Tdb,Twb)));
  }

double fTdbhw(double H,double W)
  {
  return((H-W*Hwo)/(Cpa+W*Cpg));
  }

double fRHdwb(double Tdb,double Twb)
  {
  if(Tdb<=Twb)
    return(100.);
  return(100.*fPvdwb(Tdb,Twb)/fPwsat(Tdb));
  }
```

```c
double fTdbwrh(double Twb,double Rh)
  {
  int i;
  double Tdb,Tdb1,Tdb2;
  if(Rh>=100.)
    return(Twb);
  Tdb1=Twb;
  Tdb2=fHdwb(Twb,Twb)/Cpa;
  for(i=0;i<25;i++)
    {
    Tdb=(Tdb1+Tdb2)/2.;
    if(fRHdwb(Tdb,Twb)>Rh)
      Tdb1=Tdb;
    else
      Tdb2=Tdb;
    }
  return(Tdb);
  }

double KMerkel(double Twb,double Ran,double App,double
   LG)
  { /* KaV/L from Merkel's equation using 4-point
    Chebyshev */
  int i;
  double asym,dHa,Ha,Hai,Hao,Hw,KaV,Tco,Tw;
  static double a[4]={.10,.40,.60,.90};
  static double w[4]={.25,.25,.25,.25};
  Tco=Twb+App;
  Hai=fHtwb(Twb);
  Hao=fHtwb(Tco+.9*Ran);
  asym=(Hao-Hai)/.9/Ran/Cpw;
  if(LG<0.)
    return(asym);
  if(LG>=asym)
    return(99.99);
  dHa=Ran*Cpw*LG;
  KaV=0.;
  for(i=0;i<4;i++)
    {
    Ha=Hai+a[i]*dHa;
    Tw=Tco+a[i]*Ran;
    Hw=fHtwb(Tw);
    if(Hw<=Ha)
      return(99.99);
    KaV=KaV+w[i]/(Hw-Ha);
    }
  KaV=Cpw*Ran*KaV;
  return(KaV);
```

```
  }
double KMarcel(double Twb,double Ran,double App,double
   LG)
   { /* KaV/L from Marcel Lefevre's equation using 10-
   point Lobatto */
   int i;
   double
    asym,Ha,Hai,Hao,KaV,Ta,Tai,Tao,Tao1,Tao2,Tco,Tho,Tw,W
    ,Wai,Wao,Ws;
   static double
    a[10]={.000000,.040233,.130613,.261038,.417361,.58263
    9,.738962,.869387,.959767,1.00000};
   static double
    w[10]={.011111,.066653,.112445,.146021,.163770,.16377
    0,.146021,.112445,.066653,.011111};
   Tco=Twb+App;
   Tho=Tco+Ran;
   Tai=Twb;
   Hai=fHtwb(Tai);
   Wai=fWsat(Tai);
   Tao=Tho;
   Hao=fHtwb(Tao);
   Wao=fWsat(Tao);
   asym=((Hao-Hai)/Cpw-(Wao-Wai)*(Tco-Two))/Ran;
   if(LG<=0.)
     return(asym);
   if(LG>=asym)
     return(99.99);
   Tao1=Twb;
   Tao2=Tho;
   for(i=0;i<25;i++)
      {
      Tao=(Tao1+Tao2)/2.;
      Hao=fHtwb(Tao);
      Wao=fWsat(Tao);
      if(Hao<Hai+Cpw*(LG*Ran+(Wao-Wai)*(Tco-Two)))
        Tao1=Tao;
      else
        Tao2=Tao;
      }
   KaV=0.;
   for(i=0;i<10;i++)
      {
      Ha=Hao+a[i]*(Hai-Hao);
      Ta=Ftwbh(Ha);
      W=fWsat(Ta);
      Tw=(Cpw*(LG*Tho-(Wao-W)*Two)-Hao+Ha)/Cpw/(LG-Wao+W);
      Ws=fWsat(Tw);
```

```
    if(Tw<=Ta)
      return(99.99);
    KaV=KaV+w[i]/(Lewis*Cpa*(Tw-Ta)+(Hwo+Cpg*Ta)*(Ws-
    W)/(1.+W));
    }
  KaV=Cpw*Ran*KaV;
  return(KaV);
  }

double KExact(double Tdb,double Twb,double Ran,double
    App,double LG)
  { /* KaV/L from exact equations using 4th-order Runge-
    Kutta */
  int i,j,k;
  double
    asym,H,Ha,dH[4],Hai,Hao,KaV,K1,K2,LG1,LG2,Ta,Tco,Tho,
    Tw,W,Wa,dW[4],Wai,Wao,Ws,dY;
  double A[3]={.5,.5,1.}; /* 4th-order Runge-Kutta */
  double B[4]={1./6,1./3,1./3,1./6}; /* coefficients */
  int n=10; /* number of steps to integrate */
  Tco=Twb+App; /* compute asymptote */
  Tho=Tco+Ran;
  Hai=fHdwb(Tdb,Twb);
  Wai=fWdwb(Tdb,Twb);
  Hao=fHtwb(Tho);
  Wao=fWsat(Tho);
  asym=((Hao-Hai)/Cpw-(Wao-Wai)*(Tco-Two))/Ran;
  if(LG==0.) /* for L/G=0 return asymptote */
    return(asym);
  if(LG>=asym) /* for L/G>asymptote return 99.99 */
    return(99.99);
  K1=0.; /* lower bound for KaV */
  K2=999.; /* upper bound for KaV */
  dY=1./n; /* differential step size */
  for(i=0;i<25;i++) /* bisection search method */
    {
    KaV=(K1+K2)/2;
    LG2=LG+Wai-Wao; /* exit L/G is unknown */
    Ha=Hai; /* begin integration at air inlet */
    Wa=Wai;
    for(j=0;j<n;j++) /* step upward through the fill */
      {
      for(k=0;k<4;k++)
        {
        H=Ha;
        W=Wa;
        if(k)
          {
          H+=dY*A[k-1]*dH[k-1];
```

```c
        W+=dY*A[k-1]*dW[k-1];
        }
      LG1=LG2+W-Wai; /* local L/G */
      Ta=fTdbhw(H,W); /* local Tdb */
      Tw=Two+(LG2*Cpw*(Tco-Two)+H-Hai)/Cpw/LG1; /*
  local Tw   */
      Ws=fWsat(Tw); /* local Ws   */
      if(Tw>=Tho) /* check for overshoot */
        goto overshoot;
      dW[k]=LG*KaV*(Ws-W)/(1.+W);
      dH[k]=LG*KaV*((Ws-
  W)/(1.+W)*(Hwo+Cpg*Ta)+Lewis*Cpa*(Tw-Ta));
      }
    for(k=0;k<4;k++) /* last step of Runge-Kutta */
      {
      Ha+=dY*B[k]*dH[k];
      Wa+=dY*B[k]*dW[k];
      }
    }
  Wao=Wa; /* exit humidity is now known */
  if(Tw<Tho) /* bisection comparison */
    K1=KaV;
  else
    K2=KaV;
  continue;
  overshoot: /* overshoot occurrs if KaV is too large
  */
  K2=KaV;
  }
 return(KaV);
 }
int main(int argc,char**argv,char**envp)
 {
 double App,KaV,LG,Ran,Rh,Tco,Tdb,Tho,Twb;
 printf("\n");
 printf("                  computed KaV/L <-more nearly
  exact>\n");
 printf("Twb Ran App  L/G Merkel Marcel Rh= 50 Rh= 75
  Rh=100\n");
 for(Twb=50;Twb<=60;Twb+=10)
   {
   for(Ran=20;Ran<=30;Ran+=10)
     {
     for(App=10;App<=20;App+=10)
       {
       Tco=Twb+App;
       Tho=Tco+Ran;
       for(LG=.5;LG<=2;LG+=.5)
```

```c
              {
              printf("%3.0lf %3.0lf %3.0lf
    %4.2lf",Twb,Ran,App,LG);
              KaV=KMerkel(Twb,Ran,App,LG);
              if(KaV<99)
                printf(" %6.3lf",KaV);
              else
                printf(" ******");
              KaV=KMarcel(Twb,Ran,App,LG);
              if(KaV<99)
                printf(" %6.3lf",KaV);
              else
                printf(" ******");
              for(Rh=50;Rh<=100;Rh+=25)
                {
                Tdb=fTdbwrh(Twb,Rh);
                KaV=KExact(Tdb,Twb,Ran,App,LG);
                if(KaV<99)
                  printf(" %6.3lf",KaV);
                else
                  printf(" ******");
                }
              printf("\n");
              }
            }
          }
        }
  return(0);
  }
```

The output of the program is listed below.

				computed KaV/L		<-more nearly exact>		
Twb	Ran	App	L/G	Merkel	Marcel	Rh= 50	Rh= 75	Rh=100
50	20	10	0.50	2.325	2.219	2.330	2.287	2.256
50	20	10	1.00	5.142	5.184	5.399	5.336	5.287
50	20	10	1.50	******	******	******	******	******
50	20	10	2.00	******	******	******	******	******
50	20	20	0.50	1.102	1.062	1.106	1.094	1.086
50	20	20	1.00	1.456	1.428	1.486	1.472	1.463
50	20	20	1.50	2.462	2.552	2.658	2.634	2.616
50	20	20	2.00	******	******	******	******	******
50	30	10	0.50	2.936	2.804	2.929	2.884	2.851
50	30	10	1.00	7.586	7.687	7.974	7.805	7.805
50	30	10	1.50	******	******	******	******	******
50	30	10	2.00	******	******	******	******	******
50	30	20	0.50	1.432	1.384	1.438	1.425	1.416
50	30	20	1.00	2.011	1.986	2.063	2.047	2.035
50	30	20	1.50	4.224	4.604	4.878	4.760	4.731
50	30	20	2.00	******	******	******	******	******
60	20	10	0.50	1.707	1.656	1.753	1.720	1.696

60	20	10	1.00	2.639	2.635	2.773	2.734	2.705
60	20	10	1.50	******	******	******	******	******
60	20	10	2.00	******	******	******	******	******
60	20	20	0.50	0.835	0.817	0.858	0.849	0.843
60	20	20	1.00	1.012	1.004	1.054	1.044	1.037
60	20	20	1.50	1.334	1.359	1.426	1.413	1.404
60	20	20	2.00	2.263	2.557	2.689	2.665	2.647
60	30	10	0.50	2.124	2.060	2.169	2.136	2.111
60	30	10	1.00	3.485	3.491	3.658	3.616	3.586
60	30	10	1.50	******	******	******	******	******
60	30	10	2.00	******	******	******	******	******
60	30	20	0.50	1.073	1.051	1.102	1.093	1.086
60	30	20	1.00	1.347	1.343	1.407	1.396	1.389
60	30	20	1.50	1.904	1.967	2.060	2.046	2.036
60	30	20	2.00	4.193	5.420	5.854	5.854	5.854

also by D. James Benton

3D Articulation: Using OpenGL, ISBN-9798596362480, Amazon, 2021 (book 3 in the 3D series).

3D Models in Motion Using OpenGL, ISBN-9798652987701, Amazon, 2020 (book 2 in the 3D series.

3D Rendering in Windows: How to display three-dimensional objects in Windows with and without OpenGL, ISBN-9781520339610, Amazon, 2016 (book 1 in the 3D series).

A Synergy of Short Stories: The whole may be greater than the sum of the parts, ISBN-9781520340319, Amazon, 2016.

Azeotropes: Behavior and Application, ISBN-9798609748997, Amazon, 2020.

bat-Elohim: Book 3 in the Little Star Trilogy, ISBN-9781686148682, Amazon, 2019.

Boilers: Performance and Testing, ISBN: 9798789062517, Amazon 2021.

Combined 3D Rendering Series: 3D Rendering in Windows®, 3D Models in Motion, and 3D Articulation, ISBN-9798484417032, Amazon, 2021.

Complex Variables: Practical Applications, ISBN-9781794250437, Amazon, 2019.

Compression & Encryption: Algorithms & Software, ISBN-9781081008826, Amazon, 2019.

Computational Fluid Dynamics: an Overview of Methods, ISBN-9781672393775, Amazon, 2019.

Computer Simulation of Power Systems: Programming Strategies and Practical Examples, ISBN-9781696218184, Amazon, 2019.

Contaminant Transport: A Numerical Approach, ISBN-9798461733216, Amazon, 2021.

CPUnleashed! Tapping Processor Speed, ISBN-9798421420361, Amazon, 2022.

Curve-Fitting: The Science and Art of Approximation, ISBN-9781520339542, Amazon, 2016.

Death by Tie: It was the best of ties. It was the worst of ties. It's what got him killed., ISBN-9798398745931, Amazon, 2023.

Differential Equations: Numerical Methods for Solving, ISBN-9781983004162, Amazon, 2018.

Equations of State: A Graphical Comparison, ISBN-9798843139520, Amazon, 2022.

Evaporative Cooling: The Science of Beating the Heat, ISBN-9781520913346, Amazon, 2017.

Forecasting: Extrapolation and Projection, ISBN-9798394019494, Amazon 2023.

Heat Engines: Thermodynamics, Cycles, & Performance Curves, ISBN-9798486886836, Amazon, 2021.

Heat Exchangers: Performance Prediction & Evaluation, ISBN-9781973589327, Amazon, 2017.

Heat Recovery Steam Generators: Thermal Design and Testing, ISBN-9781691029365, Amazon, 2019.
Heat Transfer: Heat Exchangers, Heat Recovery Steam Generators, & Cooling Towers, ISBN-9798487417831, Amazon, 2021.
Heat Transfer Examples: Practical Problems Solved, ISBN-9798390610763, Amazon, 2023.
The Kick-Start Murders: Visualize revenge, ISBN-9798759083375, Amazon, 2021.
Jamie2: Innocence is easily lost and cannot be restored, ISBN-9781520339375, Amazon, 2016-18.
Kyle Cooper Mysteries: Kick Start, Monte Carlo, and Waterfront Murders, ISBN-9798829365943, Amazon, 2022.
The Last Seraph: Sequel to Little Star, ISBN-9781726802253, Amazon, 2018.
Little Star: God doesn't do things the way we expect Him to. He's better than that! ISBN-9781520338903, Amazon, 2015-17.
Living Math: Seeing mathematics in every day life (and appreciating it more too), ISBN-9781520336992, Amazon, 2016.
Lost Cause: If only history could be changed..., ISBN-9781521173770, Amazon, 2017.
Mass Transfer: Diffusion & Convection, ISBN-9798702403106, Amazon, 2021.
Mill Town Destiny: The Hand of Providence brought them together to rescue the mill, the town, and each other, ISBN-9781520864679, Amazon, 2017.
Monte Carlo Murders: Who Killed Who and Why, ISBN-9798829341848, Amazon, 2022.
Monte Carlo Simulation: The Art of Random Process Characterization, ISBN-9781980577874, Amazon, 2018.
Nonlinear Equations: Numerical Methods for Solving, ISBN-9781717767318, Amazon, 2018.
Numerical Methods: Nonlinear Equations, Numerical Calculus, & Differential Equations, ISBN-9798486246845, Amazon, 2021.
Orthogonal Functions: The Many Uses of, ISBN-9781719876162, Amazon, 2018.
Overwhelming Evidence: A Pilgrimage, ISBN-9798515642211, Amazon, 2021.
Particle Tracking: Computational Strategies and Diverse Examples, ISBN-9781692512651, Amazon, 2019.
Plumes: Delineation & Transport, ISBN-9781702292771, Amazon, 2019.
Power Plant Performance Curves: for Testing and Dispatch, ISBN-9798640192698, Amazon, 2020.
Practical Linear Algebra: Principles & Software, ISBN-9798860910584, Amazon, 2023.
Props, Fans, & Pumps: Design & Performance, ISBN-9798645391195, Amazon, 2020.
Remediation: Contaminant Transport, Particle Tracking, & Plumes, ISBN-9798485651190, Amazon, 2021.

ROFL: Rolling on the Floor Laughing, ISBN-9781973300007, Amazon, 2017.
Seminole Rain: You don't choose destiny. It chooses you, ISBN-9798668502196, Amazon, 2020.
Septillionth: 1 in 10^{24}, ISBN-9798410762472, Amazon, 2022.
Software Development: Targeted Applications, ISBN-9798850653989, Amazon, 2023.
Software Recipes: Proven Tools, ISBN-9798815229556, Amazon, 2022.
Steam 2020: to 150 GPa and 6000 K, ISBN-9798634643830, Amazon, 2020.
Thermochemical Reactions: Numerical Solutions, ISBN-9781073417872, Amazon, 2019.
Thermodynamic and Transport Properties of Fluids, ISBN-9781092120845, Amazon, 2019.
Thermodynamic Cycles: Effective Modeling Strategies for Software Development, ISBN-9781070934372, Amazon, 2019.
Thermodynamics - Theory & Practice: The science of energy and power, ISBN-9781520339795, Amazon, 2016.
Version-Independent Programming: Code Development Guidelines for the Windows® Operating System, ISBN-9781520339146, Amazon, 2016.
The Waterfront Murders: As you sow, so shall you reap, ISBN-9798611314500, Amazon, 2020.
Weather Data: Where To Get It and How To Process It, ISBN-9798868037894, Amazon, 2023.

www.ingramcontent.com/pod-product-compliance
Lightning Source LLC
Chambersburg PA
CBHW031433210526
45464CB00005B/2177